C0 BYV 826

Phillips Library
Bethany College
Bethany, W. Va. 26032

Book 6

The Evolution of Reproduction

REPRODUCTION IN MAMMALS

Book 6

The Evolution of Reproduction

EDITED BY

C. R. AUSTIN
Fellow of Fitzwilliam College
Charles Darwin Professor of Animal Embryology
University of Cambridge

AND

R. V. SHORT, FRS
Director of the Medical Research Council
Unit of Reproductive Biology, and
Honorary Professor, Department of Obstetrics and Gynaecology
University of Edinburgh

ILLUSTRATIONS BY JOHN R. FULLER

CAMBRIDGE UNIVERSITY PRESS
CAMBRIDGE
LONDON · NEW YORK · MELBOURNE

Published by the Syndics of the Cambridge University Press
The Pitt Building, Trumpington Street, Cambridge CB2 1RP
Bentley House, 200 Euston Road, London NW1 2DB
32 East 57th Street, New York, NY 10022, USA
296 Beaconsfield Parade, Middle Park, Melbourne 3206, Australia

Library of Congress catalogue card number:

ISBN 0 521 21286 3 hard covers
ISBN 0 521 29085 6 paperback

First published 1976

Printed in Great Britain at the
University Printing House, Cambridge
(Euan Phillips, University Printer)

Contents

Contributors to Book 6

C. R. Austin,
Physiological Laboratory,
Downing Street,
Cambridge, CB2 3EG.

P. A. Jewell,
Department of Zoology,
Royal Holloway College,
Egham Hill, Egham,
Surrey, TW20 0EX.

S. Ohno,
Department of Biology,
City of Hope National Medical Center,
1500 East Duarte Road,
Duarte, California 91010,
USA

G. B. Sharman,
School of Biological Sciences,
Macquarie University,
North Ryde, New South Wales 2113,
Australia.

R. V. Short,
Medical Research Council Unit of Reproductive Biology,
2 Forrest Road,
Edinburgh EH1 2QW

Preface

Reproduction in Mammals is intended to meet the needs of undergraduates reading Zoology, Biology, Physiology, Medicine, Veterinary Science and Agriculture, and as a source of information for advanced students and research workers. It is published as a series of small text books dealing with all major aspects of mammalian reproduction. Each of the component books is designed to cover independently fairly distinct subdivisions of the subject, so that readers can select texts relevant to their particular interests and needs, if reluctant to purchase the whole work. The contents lists of all the books are set out on the next page.

In this new addition to the series we examine the ways in which the forces of evolution have moulded the various systems implicated in mammalian reproduction. The principal questions we ask here are: Why did mammals opt for a 1:1 sex ratio? How did viviparity evolve? Does sexual selection tend to reinforce or oppose natural selection? What reproductive mechanisms are involved in the formation and maintenance of new species? Can we see evidence of evolutionary influences even at gamete levels? Although we can find answers to some of these questions, we hope that Book 6 will generate many new questions in the reader's mind.

Books in this series

Book 1. Germ Cells and Fertilization
Primordial germ cells. T. G. Baker
Oogenesis and ovulation. T. G. Baker
Spermatogenesis and the spermatozoa. V. Monesi
Cycles and seasons. R. M. F. S. Sadleir
Fertilization. C. R. Austin

Book 2. Embryonic and Fetal Development
The embryo. Anne McLaren
Sex determination and differentiation. R. V. Short
The fetus and birth. G. C. Liggins
Manipulation of development. R. L. Gardner
Pregnancy losses and birth defects. C. R. Austin

Book 3. Hormones in Reproduction
Reproductive hormones. D. T. Baird
The hypothalamus. B. A. Cross
Role of hormones in sex cycles. R. V. Short
Role of hormones in pregnancy. R. B. Heap
Lactation and its hormonal control. Alfred T. Cowie

Book 4. Reproductive Patterns
Species differences. R. V. Short
Behavioural patterns. J. Herbert
Environmental effects. R. M. F. S. Sadleir
Immunological influences. R. G. Edwards
Aging and reproduction. C. E. Adams

Book 5. Artificial Control of Reproduction
Increasing reproductive potential in farm animals. C. Polge
Limiting human reproductive potential. D. M. Potts
Chemical methods of male contraception. Harold Jackson
Control of human development. R. G. Edwards
Reproduction and human society. R. V. Short
The ethics of manipulating reproduction in man. C. R. Austin

Book 6. The Evolution of Reproduction
The development of sexual reproduction. S. Ohno
Evolution of viviparity in mammals. G. B. Sharman
Selection for reproductive success. P. A. Jewell
The origin of species. R. V. Short
Specialization of gametes. C. R. Austin

1 The development of sexual reproduction

S. Ohno

As man values himself highly, he tends to think that any genetic system that operates in him must necessarily represent the zenith of evolutionary refinement. This surely is a naive idea. Should man argue that being 'created in God's own image' he should be able to outrun a horse and outmuscle an elephant? It seems that evolutionary successes have always been attained on the give-and-take basis of sacrificing a little here to gain a little more elsewhere, gains made in different parts of the genome being the basis of specialization and, therefore, adaptive radiation. Indeed, man has sacrificed a great deal in the process of attaining his elevated station in the animal kingdom. Being congenitally deficient in urate oxidase, he is more vulnerable to attacks of gout than most other mammals. Being incapable of internal production of vitamin C, he used to suffer during the long winters for the want of fresh fruits.

Thus, we come to realize that, while other parts of the genome have undergone considerable changes during the course of the evolution of mammals, natural selection has conserved an ancient sex-determining mechanism *in toto*, apparently because nothing better could be found. Indeed, the XX/XY chromosomal system, with occasional modifications which are more apparent than real, operates in all forms of vertebrates, as well as invertebrates, and even in some plants. In fact, the sex chromosome was first recognized by H. Henking, in 1891, in the testis of the heteropteran insect *Pyrrhocoris apterus*. At the first meiotic prophase in each nucleus of primary spermatocytes, one densely stained body stood out in sharp contrast to the fine, thread-like chromosomes. He regarded this body as a nucleolus. However, when he examined a pair of secondary spermatocytes, he noted that the body was always incorporated into one of the two daughter spermatocytes. Consequently, two kinds of haploid spermatocytes were produced in equal numbers: one with 11 chromosomes only, and the other with this body in addition to 11 chromosomes. He was no longer certain that the body was a mere nucleolus, and since he arrived at no clear conclusion, he labelled it 'X' for unknown. Eleven years were to elapse before McClung realized that, in a grasshopper, the sex of a zygote is indeed determined by whichever of the above two kinds of sperm fertilizes the egg. Nevertheless, it is in Henking's honour that today the term X-chromosome denotes the sex chromosome which is present twice in the homogametic female.

Whether it be the male heterogamety of XX/XY-type, or the female heterogamety of ZW/ZZ-type, the inevitable consequence of such a chromosomal sex-determining mechanism is the one-to-one sex ratio. Having as many males as females in a population is not necessarily a good thing, particularly for large organisms which exact heavy tolls from their environment, not only in the amount of food they consume, but also in the amount of waste they excrete. Would it not be better to reduce the proportion of essentially non-productive males? If poultry breeders and dairy farmers had their way, they would certainly prefer a very biassed sex ratio in favour of females. Should not Nature have preferred the same? In order to get a better perspective of the subject, we shall first consider various reproductive alternatives to the one-to-one sex ratio that are being employed by various lower vertebrates. It is perhaps remarkable that although vertebrates must have experimented with different means to escape from the stranglehold imposed by the one-to-one sex ratio, this antique sex-determining mechanism is still with us.

ALTERNATIVE METHODS OF REPRODUCTION

Synchronous hermaphroditism

Natural selection is most effective when it operates among individuals of the same species which differ in their genetic constitution. It has often been pointed out that the degree of hereditary variability within a species is greatly enhanced by cross-fertilization – hence the development of sexual reproduction and a chromosomal sex-determining mechanism. But such an argument is not altogether sound.

It is believed that at the beginning of Cambrian times, some 500 million years ago, the original ancestors of the vertebrates appeared; they were uncomplicated creatures, simple sessile animals which resembled tiny, deep-sea forms of the present day Ascidiacea, the sea squirts or tunicates. Although placed in the Phylum Chordata, these animals, attached by a stalk to the ocean bottom, consist of little except a digestive tract. Long cilia in the 'pharynx' waft food particles drifting past into the 'stomach'. As sessile ciliary feeders the tunicates have reached an evolutionary cul-de-sac. But, what were the characteristics of the larval forms of those early creatures? In the free-swimming tadpole-like larvae of modern tunicates there is a muscular swimming tail, strengthened by a stout but flexible notochord, predecessor of the vertebral column. There exists a longitudinal dorsal nerve cord which, in the head region, receives sensory information from rudimentary sense organs. It is from the larva of a tunicate-like form of ancient times that the body form of the first true vertebrate sprang. In evolution this process of paedomorphosis happened time and again: the adult stage was eliminated, and the larval form became sexually mature and reproduced.

Tunicates of today are synchronous hermaphrodites, having a mature testis and ovary at the same time. But self-fertilization is avoided by a simple anatomical device, a genital duct filled with one type of gamete (sperms or eggs) which shuts off the other genital tract. Thus, the sea squirt ejects eggs or sperms, but never both at the same time. Indeed, using electrophoretic variants of a few enzymes, we found that *Ciona intestinalis* of the Southern

California coast is as genetically polymorphic as any other species of bisexual vertebrates. Hermaphroditism is generally associated with cross-fertilization.

The fact that modern tunicates are synchronous hermaphrodites, however, should not be taken as evidence that the ancestral vertebrate stocks started with this mode of reproduction, for gonochorism (separation of the sexes) is a practice that prevails throughout the animal kingdom. Synchronous hermaphroditism is a deviation from the norm that is employed sporadically but recurrently by various lower vertebrates. Thus, we see this mode of reproduction in certain of the advanced teleost fish, while the more primitive sharks, rays and sturgeons uniformly remain faithful gonochorists. Synchronous hermaphroditism is the normal mode of reproduction in many species of sea basses of the Family Serranidae; the gonads consist of paired ovotestes fused posteriorly, thus the cavities of the two lobes open into a common duct. Self-fertilization is a possibility, but their group-spawning behaviour fortunately ensures cross-fertilization most of the time (Fig. 1-1).

Of course, the ability to self-fertilize can come in handy. The depths of the great oceans are eternally dark and the food supply is scarce. For fish that live under these conditions, the self-fertilizing kind of synchronous hermaphroditism might be a decisive advantage. Small fish of the Order Cyprinodontiformes (toothed carps) inhabit freshwater ponds which are subject to seasonal fluctuation in water levels, and an individual may find itself alone in a small water hole if there is a drought. Again, self-fertilization becomes the only means of propagation. Indeed, one egg-bearing cyprinodont, *Rivulus marmoratus*, is an internally self-fertilizing synchronous hermaphrodite; its population constitutes a genetically homozygous clone as judged from acceptance of reciprocal fin grafts among members.

Needless to say, such self-fertilizing systems can be beneficial only as an interim measure; prolonged self-indulgence would surely doom the species to extinction. On the other hand, there are no obvious shortcomings with the cross-fertilizing of synchronous hermaphroditism.

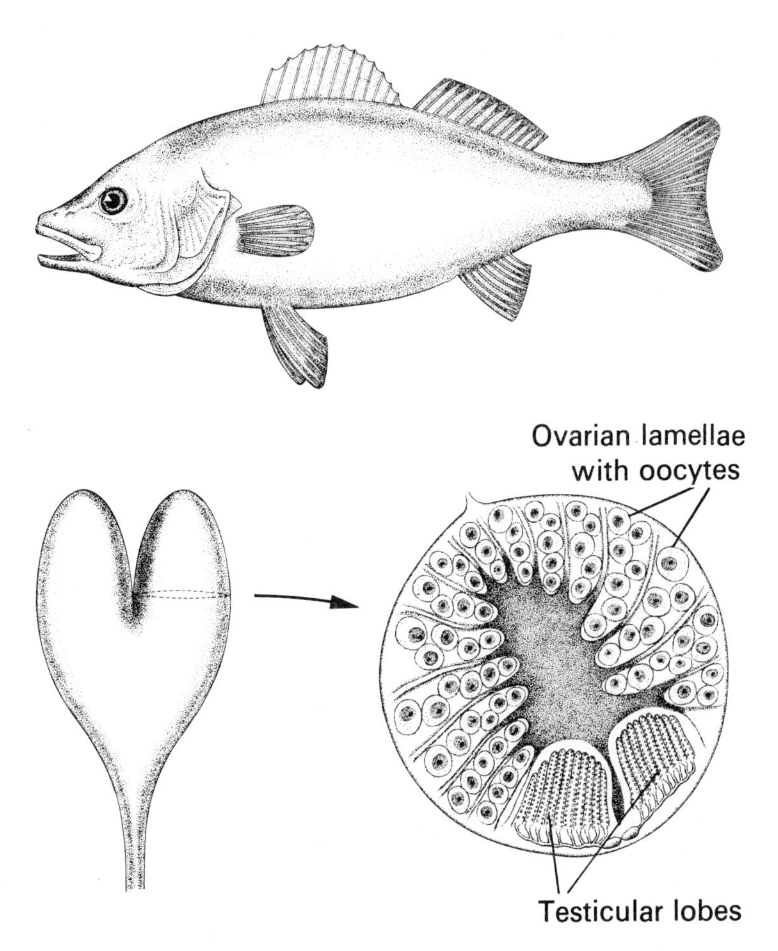

Fig. 1-1. Sea basses (the Family Serranidae) are synchronous hermaphrodites. The gross appearance of a pair of gonads is shown at the bottom left, and a cross-section of a lobe at the bottom right. The large central cavity surrounded by ovarian lamellae filled with oocytes constitutes the oviduct. Two small testicular lobes contain seminiferous tubules which collectively open into a pair of sperm ducts. (Modified from U. D'Ancona. *Nova Thalassia* **1**, 15 (1949).)

Asynchronous hermaphroditism

At all stages in the gonadal development of synchronous hermaphrodites, the preponderance of ovarian over testicular tissue is evident. What if testicular development in the hermaphroditic gonad becomes suppressed until later in life? The individual would then be classified as an asynchronous hermaphrodite of the protogynous type, in that all individuals of a population would function as females when they were young and switch to being males when they were aged. This seems an ideal way of escaping from the burden imposed by the one-to-one sex ratio inherent in gonochorism, since progressive mortality would surely reduce the number of males in a population to a level substantially below 50 per cent. Furthermore, those that did manage to transform to males would represent the cream of the population. Protogynous asynchronous hermaphroditism is practised by groupers and related species belonging to the Family Serranidae, which are perch-like fish; it will be recalled that the sea basses, which belong to the same family, are synchronous hermaphrodites. This form of asynchronous hermaphroditism is also practised by swamp eels of the Order Synbranchiformes.

Asynchronous hermaphroditism can also be of the protandrous type when the testis develops before the ovary. This would appear to ensure a preponderance of males in the population, but nevertheless it is practised successfully by a wide variety of teleosts such as porgies and sea breams.

True parthenogenesis and gynogenic parthenogenesis

One can purchase the Amazon molly (*Poecilia formosa*) from almost any pet shop that has a sizeable aquarium section, for this small, ovoviviparous toothed carp, from Texas and Mexico, has a rather pleasing coloration. It would be impossible, however, to buy a male Amazon molly, for this is, as the name implies, an all-female gynogenic parthenogen. The species apparently arose as an interspecific hybrid between *P. latipinna* and *P. mexicana*, and it exists as a 'sexual parasite' of one of the parental species, the sailfin molly, *P. latipinna*. Virgin Amazon mollies do not produce viable offspring; they will only do so after copulating with a male sailfin molly. The male, however, does not contribute genetically to the genome of *P. formosa* at this mating; his sperms merely serve to induce parthenogenetic development in the diploid egg of the Amazon molly. A similar, but more complicated situation, has been found in another toothed carp interspecific hybrid, *Poecilliopsis monacha-lucida*, inhabiting rivers of Northern Mexico.

Among reptiles, parthenogens have been found in whiptail lizards of the Genus *Cnemidophorus*, which inhabit arid regions of Arizona and New Mexico. In geographical boundaries between neighbouring bisexual species, interspecific hybrids may arise. True parthenogenesis and gynogenic parthenogenesis are the means employed by these hybrids to escape sterility and establish themselves as all-female species. Being of hybrid origin, these all-female species could escape the deleterious effect of inbreeding for a long time, in spite of their parthenogenic mode of reproduction.

If a large number of chicken or turkey eggs from virgin hens are placed in an incubator, a few will start spontaneous embryonic development, and an even smaller percentage will hatch and even grow up to be adult parthenogens. It is of interest to note that infection with oncogenic viruses seems to enhance such parthenogenic development in birds.

Unfortunately for chicken farmers, in birds the female is the heterogametic (ZW) sex. Since it is the fusion of two haploid chromosome sets following the second meiotic division that starts parthenogenic development, the sex chromosome constitution of parthenogens is either ZZ (♂) or WW, and the latter is lethal in birds. Thus, all surviving chicken or turkey parthenogens are males. The benefit that could be derived from the establishment of an all-female strain of an egg-producing chicken is obvious. Furthermore, the fact that in birds parthenogens arise from the fusion of two haploid cells (oocyte and polar body) rather than from chromosome duplication in one haploid oocyte, insures that parthenogens would long escape the deleterious effect of being homozygous at too many gene loci.

In all-female parthenogenic species of both fish and lizards, triploid clones, having a diploid set of one parental chromosome complement and a haploid set of the other, have been found. It appears that triploidy has a stabilizing influence on the parthenogenic state. While triploidy is an apparently lethal condition in mammals, with a significant proportion of spontaneously aborted human fetuses being triploids, this is not so in birds. Following our discovery in 1963 of an adult triploid chicken of ZZW sex chromosome constitution, more adult triploid chickens have now been reported. Unfortunately, ZZW triploids are intersexes with ovotestes. Maybe ZWW triploids could be normal females; if so, it might be possible to create an all-female triploid parthenogenic clone from them. Hope springs eternal!

WHY A ONE-TO-ONE SEX RATIO AND WHY SEX CHROMOSOMES?

Justification of the chromosomal sex-determining mechanism

The foregoing discussions have demonstrated the point that, in order to generate sufficient genetic diversity, it is not necessary to have irreversibly determined genetic males and females in a population. Asynchronous hermaphroditism and even synchronous hermaphroditism, if an anatomical or behavioural barrier is added, is just as effective in generating genetic diversity. Furthermore, even some types of parthenogenesis can maintain sufficient genetic heterozygosity within a population for long periods of time. When faced with a confined living space and a limited food supply, the shortcomings of an all-female parthenogenic mode of reproduction are more than amply compensated for by not having to maintain essentially non-productive males in the population. Indeed, during the course of evolution, venturesome species have experimented time and again with doing away with the chromosomal sex-determining mechanism.

Why then have mammals been uniformly left with the XX/XY mechanism? Mammals in general are of large size, so that the potential benefit to be gained from not having to maintain numerous males in the population should be enormous. The argument that the mammalian body cavity is not large enough to house both sets of reproductive organs does not carry much conviction, for the total space occupied by most male reproductive tracts is minuscule. Perhaps mammals, like most other vertebrates and invertebrates, have been unable to change their method of sex-determination and have simply been making the best of it. Critical minds, however, cannot be satisfied with such an explanation.

It is a curious fact that, in the plant kingdom, the sex chromosomes and sexual reproduction are most often found among members of the primitive Phylum Bryophyta. Many of the mosses have an XX/XY scheme of sex determination. Some readers may be surprised to learn that the size difference between the large X- and the small Y-chromosome of mosses, such as *Sphaerocarpus donnellii*, is even more pronounced than that which we usually see between the mammalian X- and Y-chromosomes (Fig. 1-2). On the other hand, 'modern' plants with beautiful blooms are, as a rule, hermaphrodites, the one flower bearing pollen on the stamens and ovules in the pistils. Needless to say, bees and butterflies customarily ensure that cross-fertilization rather than self-fertilization occurs. It seems that, in the plant kingdom, Nature began progressively to favour the hermaphroditic mode of reproduction.

Why then the difference in evolutionary preferences between animals and plants? Higher plants are stationary beings growing in places where fate has planted their seeds. Individual interactions are minimal in plants. Thus, having no need, they have not developed a nervous system. Members of the animal kingdom, in sharp contrast, move about restlessly; they bump into one another, and inevitably either conflict or co-operation develops between individuals with regard to territory, food supply, etc. Thus, the evolution of animals culminated in the progressive

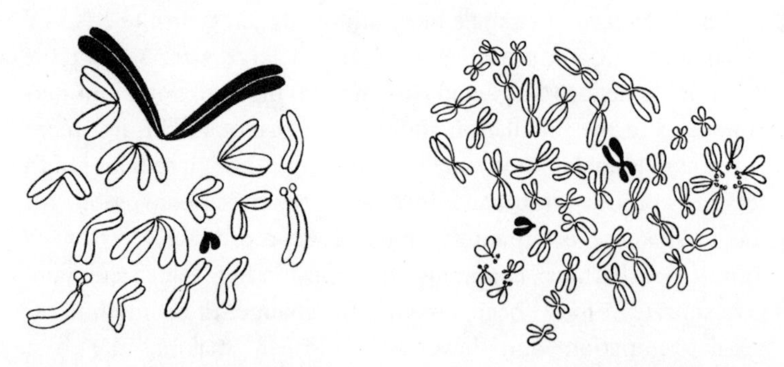

Fig. 1-2. Schematic drawings (not to scale) of the large X-chromosome and the small Y-chromosome (solid black) of a moss, *Sphaerocarpus donnellii* (2n = 16), on the left, compared with those of man, *Homo sapiens* (2n = 46), on the right.

development of the central nervous system. The way animals behave must have come under particularly strong surveillance by natural selection, for the rise and fall of an animal species must depend very heavily upon its members' behavioural patterns and social organizations. Herein may lie the *raison d'être* of the chromosomal sex-determining mechanism and its prevalence throughout the entire animal kingdom. Would salmon swim hundreds of miles up-river to the place of their birth if there were no sex at the end of the journey? If sticklebacks were hermaphroditic, would they show the very unfishlike behaviour of nest building, guarding the eggs and rearing the young, which is all performed by the male?

The nature and number of sex-determining genes

Because the sex chromosomes were recognized very early under the microscope, while the science of genetics was still in its infancy, the vague notion still persists that the whole of both X- and Y-chromosomes must necessarily be involved in the act of sex determination. A glance at Table 1-1 (p. 21) should dispel such a notion once and for all. The fact is that all except one of the known mammalian X-linked genes listed in Table 1-1 have nothing whatsoever to do with sex as such. For example, G-6-PD (glucose-6-phosphate dehydrogenase) is the first enzyme of the pentose phosphate shunt of carbohydrate metabolism. This is a ubiquitous metabolic pathway both in ontogeny and phylogeny, being already present in bacteria. Indeed, in avian species, G-6-PD is autosomally inherited. Similarly, PGK (phosphoglycerate kinase) is another ubiquitous enzyme concerned with basic glycolysis, where HGPRT (hypoxanthine–guanine phosphoribosyl transferase) is a mere purine scavenging enzyme. To ascribe any kind of sex-related role to such enzymes would be absurd. It is merely a quirk of evolutionary fate that some of the enzyme genes for daily household chores of cells have been placed on the mammalian X-chromosome. The same can be said of two genes which apparently specify blood clotting factors (haemophilia A and B are X-linked in man and dog), as well as two X-linked genes for trichromatic vision (two kinds of colourblindness are X-linked in man).

There is only one known X-linked gene that has a direct bearing on sexual development; that is the *Tfm* (*testicular feminization*) locus. The *Tfm* mutation has long been known in man and other mammals, but its X-linkage has recently been unequivocally established in the mouse by Mary Lyon. Even more recently, the X-linkage of the *Tfm* locus has been confirmed in man. As discussed by Roger Short in Book 2, Chapter 2 of this series, mammalian embryos, regardless of their sex chromosome constitutions, have an inherent tendency to develop the female phenotype; the mammalian male is essentially a female that has been exposed to androgenic steroid hormones. The action of an androgenic steroid on its target cells is mediated by the so-called nuclear-cytosol androgen-receptor protein which every androgen target cell possesses, and this receptor protein has all the qualifications of a regulatory protein. Whenever it is saturated with androgen, androgen-bound receptor molecules move into the nucleus and associate with the chromatin, thereby eliciting the induction of a target-organ-specific set of structural gene products. The *Tfm* mutation, which gives affected androgen-producing XY individuals a female phenotype, is caused by a functional deficiency of this nuclear-cytosol androgen-receptor protein. One might say that the sexual organ of the mammalian X-chromosome is its *Tfm* locus (Fig. 1-3).

Although it now appears probable that some of the enzymes for synthesis of steroid sex hormones are also X-linked, there is no reason to believe that such X-linkages are any more significant than those of G-6-PD, PGK or HGPRT. The scheme of sex hormone synthesis is such that a female gestagen, progesterone, is a precursor of androgen, while androgen in turn serves as a precursor of another female hormone, oestrogen. Accordingly, both the male and the female must necessarily possess the complete set of sex hormone synthesizing enzymes, differential

---

Fig. 1-3. Two genetic regulatory systems that determine mammalian sex. Embryonic indifferent gonads of mammals have an inherent tendency to develop toward the ovary and the rest of the mammalian body also has an inherent tendency to develop toward the female. For an embryo to develop as a male, a gene or genes on the Y-chromosome divert the developmental direction of an indifferent gonad toward the testis. The testis thus formed then produces androgenic steroid hormones, mainly testosterone.

Cells of androgen target organs such as the embryonic Wolffian duct and the urogenital sinus respond to testicular androgens, and male-specific organs such as the seminal vesicles, prostate and penis are formed. Testosterone target cells in the female are normally not exposed to androgens. Thus, they either wither away during development, or remain quiescent. The androgen responsiveness of target cells is due to the nuclear-cytosol androgen-receptor protein which is apparently specified by the X-linked *Tfm* locus. This androgen-receptor protein is present in equal amounts in both sexes. Thus, the XX female body can readily respond to externally administered androgen. The receptor protein in the cytoplasm, when saturated with androgen, acquires the acceptor binding site and moves into the nucleus to bind with the chromatin via the acceptor binding site. This binding causes the induction of a specific set of enzymes and other proteins and also increases the production of ribosomal RNA. In many target cells, testosterone is readily converted to 5α-dihydrotestosterone, and the receptor protein indeed shows a higher binding affinity for 5α-dihydrotestosterone than for testosterone. However, this conversion is not absolutely essential, for in the absence of 5α-dihydrotestosterone the receptor protein can make do with testosterone.

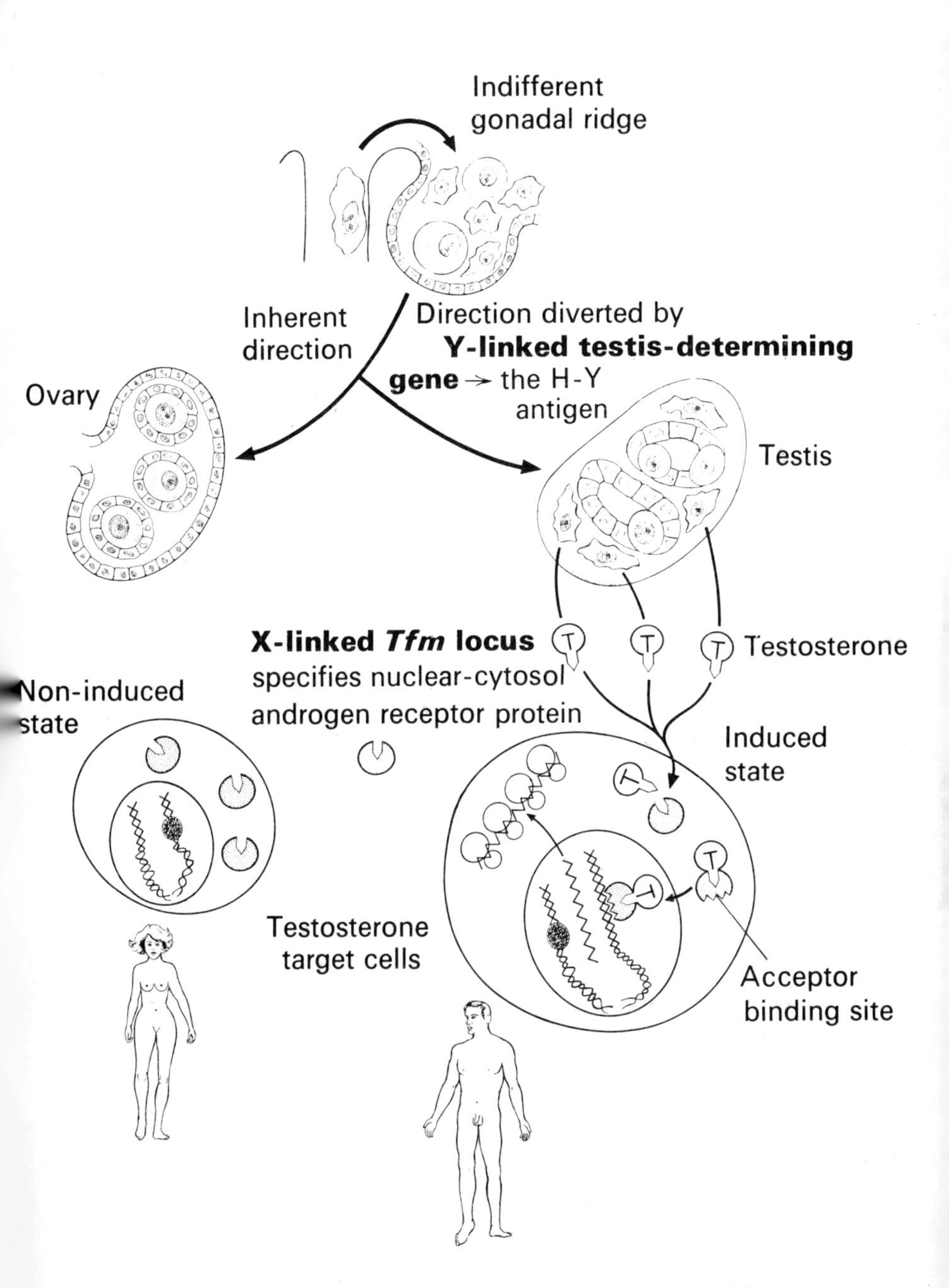
Indifferent
gonadal ridge
Inherent
direction
Direction diverted by
Y-linked testis-determining
gene → the H-Y
antigen
Ovary
Testis
X-linked *Tfm* locus
specifies nuclear-cytosol
androgen receptor protein
T
Testosterone
Non-induced
state
Induced
state
Testosterone
target cells
Acceptor
binding site

regulation of the same set of enzymes being the cause of sex differences in hormone production. No doubt, most of the enzymes of this biosynthetic pathway will be shown to be autosomally inherited. The recently demonstrated autosomal inheritance in man of a steroid-metabolizing enzyme can be explained along the same lines.

In most of the androgen target cells, but not all, testosterone is converted to 5α-dihydrotestosterone by 5α-reductase. As the androgen-receptor protein mentioned above shows a higher binding affinity for 5α-dihydrotestosterone than testosterone, 5α-reductase seems to be a rather important enzyme for the masculinization of mammalian embryos. However, this does not necessitate that it should be X-linked.

Compared to the X-chromosome, the mammalian Y-chromosome is usually a minute element, its share of the total genetic material being only 1–3 per cent of the haploid complement. Yet its minuteness is only relative, for even a tiny Y-chromosome contains at least 3×10^7 DNA base pairs. This is still an enormous amount of DNA when one realizes that it is roughly ten times the entire genetic apparatus of *E. coli*, and yet the *E. coli* genome specifies at least 2000 genes. The mammalian Y-chromosome is not only small, it also appears to be largely a dummy. For example, most of the long arm of the acrocentric human Y-chromosome which fluoresces brightly with a quinacrine mustard dye, appears dispensable, consisting of repetitious DNA base sequences of doubtful functional significance. As long as a minute region around the centromere is present, the human Y-chromosome can apparently fulfil its assigned function of inducing testicular development in undifferentiated embryonic gonads. Indeed, a number of primate species possess an extraordinarily minute Y-chromosome which corresponds only to the pericentric region of the human Y-chromosome.

Could it be that the mammalian Y-chromosome contains only one important gene, that for testicular development? (See Fig. 1-3.) Indeed, an autosomal dominant *Sxr* (*sex reversal*) gene in the mouse, discovered by Bruce Cattanach, can apparently cause testicular development in the absence of the Y, transforming genetic females into phenotypic males with testes. Such XX, *Sxr*/+'males' are positive for the Y-linked histocompatibility (H-Y) antigen, which suggests that the *Sxr* gene could be located on a microscopically invisible piece of the mouse Y-chromosome that has been translocated to an autosome. However, there are no germ cells present in the testis of these XX, *Sxr*/+'males', since the presence of two X-chromosomes dictates that the germ cells will behave like oogonia, which cannot survive in a testicular environment. In contrast, the XO, *Sxr*/+ 'male' does produce spermatozoa, although they seem to be abnormal, suggesting that other parts of the Y-chromosome may be necessary for spermatogenesis.

While we can only deduce the presence of testicular development genes on the mammalian Y, the Y-linked histocompatibility (H-Y) antigen gene is real, in that its gene product can be identified by a specific antibody. Within a highly inbred strain of mice, reciprocal skin grafts should be accepted, as individual members are virtually identical with respect to their autosomally inherited histocompatibility antigens. Surprisingly, in 1955 it was found that females reject male skin grafts, whereas skin grafts exchanged between all other sex combinations are accepted; for example, XX skin is accepted by XY males, or XO skin by XX females. Using a specific antibody raised in female mice against grafted male cells of the same strain, Boyse and his group have recently demonstrated the remarkable evolutionary conservation of this H-Y antigen. As far as the discriminatory power of his specific mouse antibody can tell us, the H-Y antigen of the mouse is either strongly cross-reactive, or identical with an antigen found in male humans, rats, guinea pigs and rabbits. Furthermore, among amphibians, the H-Y antigen is found in male leopard frogs, *Rana pipiens*, where the XX/XY scheme is known to operate, whereas the antigen is only found in females of the South African clawed toad *Xenopus laevis* which has a ZW/ZZ sex-determining mechanism. The H-Y antigen is therefore not so much Y-linked as heterogametic sex specific. Indeed, among chickens, hens (ZW) were H-Y positive, while cocks (ZZ) were negative. Such remarkable evolutionary conservation is a very strong indication of the H-Y antigen's functional importance. Accordingly, it appears quite conceivable that this H-Y antigen gene could be one and the same as the long-sought dominant sex-determining gene which induces the embryological development of heterogametic gonads. The histocompatibility antigen by definition denotes a cell surface protein, and nobody would quibble with the importance of cell surface recognition in ontogenic development.

Why is there a size discrepancy between the X- and the Y-chromosome?

The foregoing discussion has shown that neither the whole X nor the whole Y, but only very few genes, maybe only one or two, on each of these chromosomes are actually responsible for the acts of sex determination and differentiation. That being the case, the X and Y could have remained homologous, and the same size as each other. It is worth remembering that, with regard to any gene locus, a mating between the homozygote and the heterozygote unfailingly perpetuates the same two genotypes in a one-to-one ratio, provided of course that only two alleles are involved. Thus, the minimal requirement for establishing the male heterogamety type of genetic sex-determining mechanism is to have two alleles at a single gene locus, homozygotes for the recessive allele developing as females, and heterozygotes for the dominant allele developing as males.

Indeed, in many gonochorist species of fish, amphibians and reptiles, the X and the Y, or the Z and the W, are not only morphologically identical, but also largely genetically homologous. In the absence of microscopically recognizable sex chromosomes, male heterogamety or female heterogamety can be easily ascertained by sex reversal experiments.

These become particularly elegant if proper sex-linked marker genes are utilized. Although many sex reversal experiments have been done on a considerable number of fish and amphibian species, perhaps the neatest was carried out by Yamamoto in a small Japanese cyprinodont fish, *Oryzias latipes.* In this species, one of the body-coloration gene loci is closely linked to the sex-determining locus or loci. A recessive r (white) allele of the body-coloration locus is normally coupled with a female-determining gene on the X, while a dominant R (orange-red) allele is coupled with a male-determining gene on the Y. Thus, a stock can be constructed in which all the females are white (X^rX^r) and all the males are orange-red (X^rY^R). Because of the close linkage, cross-over orange-red females (X^rX^R) and white males (X^rY^r) are exceedingly rare (0.2 per cent). Now, if an appropriate amount of androgen is added to the aquarium water of larval fish, all the white X^rX^r females will be sex-reversed to functional males; subsequent mating between such X^rX^r males and normal X^rX^r females naturally produces all white females, thus confirming the female homogamety. Conversely, an addition of oestrone produces sex-reversed X^rY^R orange-red females. By mating such sex-reversed females to normal orange-red males, a three-to-one sex ratio is obtained in the progeny: Y^RY^R♂ : X^rY^R♂ : X^rX^r♀ = 1 : 2 : 1. If Y^RY^R♂ thus obtained are then mated to normal white X^rX^r females, nothing but normal X^rY^R orange-red males result, thus confirming the male heterogamety. The above example in *O. latipes* confirms that genetic sex determination can be carried out by a pair or a few pairs of alleles. Thus the loss of homology between the X and Y cannot be a necessary prerequisite for this chromosomal sex-determining mechanism (Fig. 1-4).

Why then is there such a pronounced size difference between the mammalian X- and Y-chromosomes? There is no doubt that the mammalian Y is not the genetic equal of the mammalian X. While the XO condition is present in a variety of mammals and XO mice are even fertile, the OY condition is uniformly lethal in mammals. Such genetic degeneracy of the Y-chromosome does not necessarily represent a modern, evolutionary refinement. It should be recalled that a pronouncedly heteromorphic XY pair is

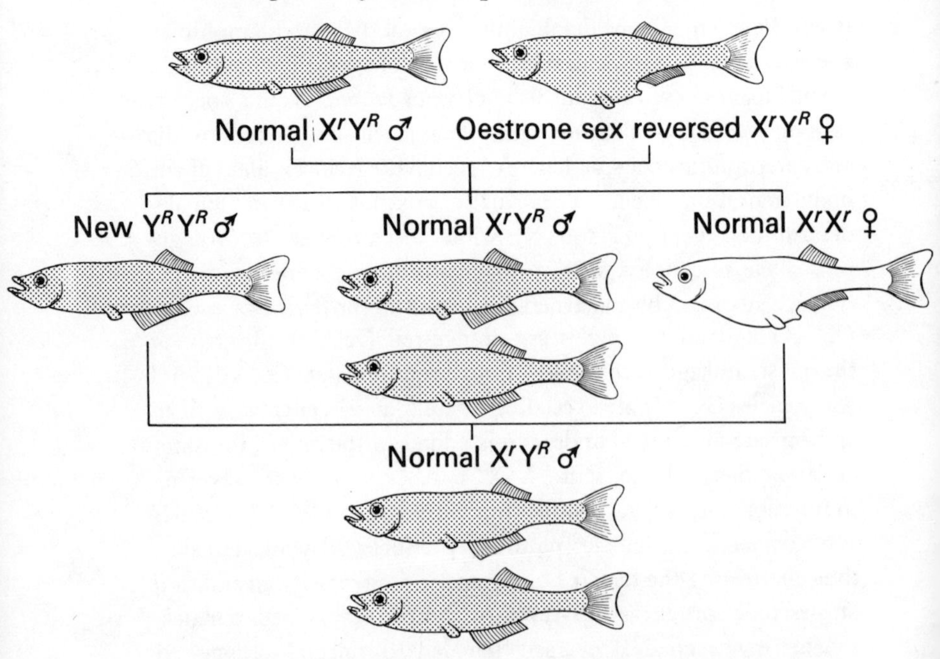

Fig. 1-4. The XX/XY type of chromosomal sex-determining mechanism has been beautifully demonstrated in a small Japanese cyprinodont fish, *Oryzias latipes*, by the combined use of steroid-induced sex reversal and tightly sex-linked body coloration genes. For explanation see text.

a common occurrence among primitive plants such as mosses, and such examples abound in insects and other invertebrates. In addition, very heteromorphic XY or ZW pairs have been reported in diverse species of teleost fish, even though some of these species belong to families or orders which also contain synchronously hermaphroditic species. Such findings in teleost fish indicate that the genetic degeneration of the Y or the W can occur very rapidly in the evolutionary time scale, and does not require elaborate preparative steps.

Is there any advantage in having a degenerate or very specialized Y-chromosome, which would inevitably make the heterogametic male sex more genetically vulnerable than the homogametic female sex since almost all of the male's X-linked genes would be exposed in the hemizygous state? Perhaps such vulnerability serves to eliminate surplus males from the population; but if that were the object if would be absurd to choose the female as the heterogametic sex, as birds and many other vertebrates have done. Conventional explanations for the virtue of having a degenerate or very specialized Y-chromosome, including my own of some years ago, are too contrived to be convincing. At the risk of contriving once more, I will offer yet another. A quirk of evolutionary fate ordained that the mammalian X-chromosome had a disproportionately large share of genes with regulatory functions. As regulatory gene products need to be produced in relatively minute quantities, it was advantageous for the X-chromosome to remain forever in the functionally haploid state – in males naturally, and in female somatic cells by the X-inactivation process.

Recently, Mary Lyon has proposed that the heteromorphism between the X and the Y of mammals might have arisen by a transfer of the bulk of genetic material from the Y to the X. Since at that ancestral stage the Y would have been still largely homologous with the X, such a transfer would have effectively duplicated the X-chromosome (Fig. 1-5). Indeed, there are some indications of internal duplication within the X. For example, both anti-haemophilic factors VIII and IX are X-linked (Table 1-1) and two forms of X-linked colour blindness are known in man. However, until the amino acid sequences of their direct gene products are compared, we will not really know whether these apparent gene pairs constitute pairs of duplicated genes. On the negative side, there is only a single X-linked gene locus for each of the enzymes listed in Table 1-1. Furthermore, in the case of glucose-6-phosphate dehydrogenase, phosphoglycerate kinase and phosphorylase kinase, their apparent isozymes are autosomally inherited.

While such initial duplication nicely explains the subsequent development of the X-inactivation mechanism in mammalian females, we really have no explanation for the widespread occurrence of pronounced heteromorphism between the X and the Y

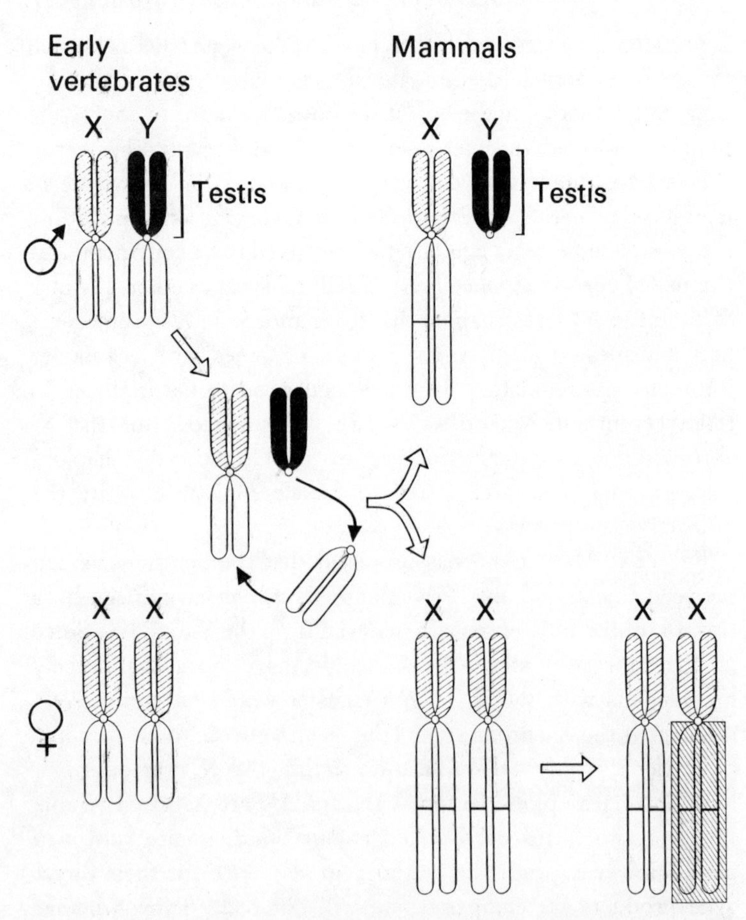

Fig. 1-5. Mary Lyon's proposal for the origin of mammalian XY heteromorphism. At the pre-mammalian ancestral stage, the X and the Y were morphologically identical (the situation above on the left). However, a differential segment on the Y which contains a testis-determining gene or genes (solid black) avoids genetic crossing-over with its counterpart on the X (hatched). Free cross-overs were permitted between the remaining homologous segment from the Y and promoted its subsequent transfer to the X. Thus, the Y became small, and the X has effectively doubled. This doubling of the X in mammals (the situation on the right) necessitated the subsequent invention of a peculiar dosage compensation mechanism which inactivates one or the other X in all somatic cells. (After M. F. Lyon. *Proceedings of the Royal Society of London, Series B* **187**, 243 (1974).)

TABLE I-I. Evolutionary conservation of mammalian X-chromosomes reflected in the homology of X-linked genes

Human X-linked gene product or disease	X-linked in other mammals
Glucose-6-phosphate dehydrogenase (of pentose phosphate shunt)	Chimpanzee, horse, donkey, hare, mouse, kangaroo
Phosphoglycerate kinase (for glycolysis)	Horse, mouse, hamster, kangaroo
Phosphorylase kinase (for glycogen utilization)	Mouse
Hypoxanthine–guanine phosphoribosyl transferase (for purine re-utilization)	Horse, mouse
Anti-haemophilic factor VIII (for blood coagulation)	Dog
Anti-haemophilic factor IX (for blood coagulation)	Dog
X_g erythrocyte antigen	Gibbon
Copper transport deficiency (Menkes kinky-hair syndrome)	Mouse
Lysyl oxidase (for collagen cross-linking)	Mouse
Dominant hypophosphataemic rickets	Mouse
Anhidrotic ectodermal dysplasia	Cattle
Nuclear-cytosol androgen-receptor protein (testicular feminization syndrome)	Mouse

The X-linkages listed have been established, not only by classical pedigree studies, but also on the following three sets of data obtained by modern techniques. (1) When inter-specific somatic cell hybrids are made *in vitro*, progressive elimination of one parental species' chromosomes tends to occur. Thus, the situation can be created in which one parental species is represented only by an X-chromosome in a somatic hybrid. In this condition, any species-specific enzyme or protein that remains in a hybrid can be considered X-linked. (2) Because of the random X-inactivation in female somatic cells, the demonstration of two hemizygous clones in the body of obligatory heterozygotes can be considered as proof of that trait's X-linkage. (3) The principle of X-inactivation does not apply to female germ cells, particularly maturing oocytes. Thus, a 2:1 ratio in enzyme activities between XX oocytes and XO oocytes strongly suggests the X-linkage of the enzymes measured.

or the Z and the W in lower animals and plants in which the X-inactivation mechanism is totally absent.

PRINCIPLE OF EVOLUTIONARY CONSERVATION

Happy were the days when all that evolutionists and geneticists had to deal with were the external appearances of organisms. In that carefree past it appeared as though natural selection were truly omnipotent in moulding the shapes and sizes of organisms to fit their particular environment.

With the advent of molecular biology, we became wise, and sadly realized the limited power of natural selection. All the living organisms on this earth, from a simplistic bacteriophage to man, utilize an identical set of triplet codons to translate the base sequence of messenger RNA to the amino acid sequence of a polypeptide chain. A codon AUG universally specifies methionine, and UUU, phenylalanine. The codons are universal, not because the triplet coding system represents an ideal solution to the problem, but merely because the first organism, or rather the first self-replicating nucleic acid, which emerged aeons ago on this earth, happened to hit upon this coding system which was reasonably efficient. Once the coding system was established, at the very beginning of life, there was no choice but to conserve it *in toto* in all the myriad descendants of that first creature. Any subsequent attempt to change the coding system would necessarily have made a mockery of all previous messages that were encoded within the DNA, thus resulting in an immediate extermination of any organism which dared to attempt a change. In this manner, all organisms are bound to the past. As they evolve, their past history becomes an increasing burden which progressively restricts their future evolutionary possibilities. As hard as modern man strives to be free he is a slave chained to the past.

With regard to any individual gene product, the cardinal rule of evolution appears to have been that the amino acid sequence of that part of the molecule which represents the active site is rigidly conserved, although elsewhere evolutionary amino acid substitutions are permissible on functionally trivial parts of the same molecule. In the case of an enzyme, the active site is that part which recognizes the appropriate substrate or coenzyme. Thus, any mutational alteration of the active site would render a mutant enzyme delinquent in the performance of its assigned function. Needless to say, such mutations have seldom been allowed to accompany a successful speciation, unless the enzyme became redundant in that species. Thus, in a molecule whose entire length represents a functionally active site, the trend for evolutionary conservation becomes extreme. Witness the fact that histone IV (with 110 amino acid residues) of cattle and of garden peas differ from each other only by two amino acid substitutions. In the histone IV gene of eukaryotes, no meaningful change has occurred in a few billion years.

This evolutionary conservation appears to extend to the regulatory elements of the genomic DNA as well. When a human diploid fibroblast is hybridized to an aneuploid rodent cell derived from liver, say a rat hepatoma cell, one often sees an awakening of liver-specific functions in the human genome part of the resultant somatic cell hybrid. Apparently, a rat regulatory molecule, which controls a liver-specific set of rat structural genes, can also recognize a liver-specific set of human structural genes and turn them on.

Thus it is no surprise to find that the X- and Y- chromosomes of divergent mammalian species have not essentially changed in spite of 125 million years of adaptive radiation. Some years ago, we proposed that the mammalian X-chromosome has been conserved *in toto*. Consequently, any gene which has been found to be X-linked in man would automatically be X-linked in all other mammalian species. Table 1-1 and Fig. 1-6 show that examples of homology are now sufficiently numerous to confirm this postulate. In the case of the Y-chromosome, the remarkable evolutionary conservation of the H-Y antigen which we already discussed, also indicates that the functionally critical part has been rigidly conserved, and that there may even be a homology between the Y and the W of vertebrates.

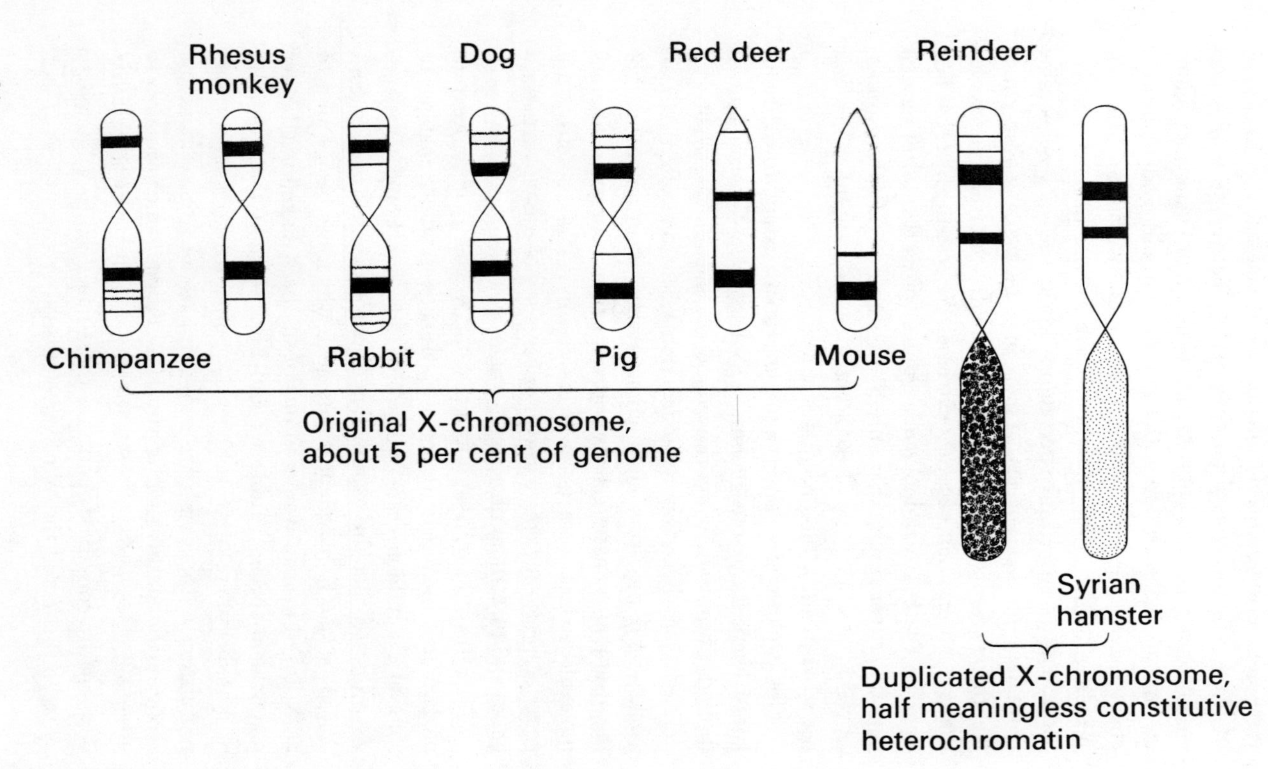
Rhesus monkey
Dog
Red deer
Reindeer
Chimpanzee
Rabbit
Pig
Mouse
Original X-chromosome,
about 5 per cent of genome
Syrian hamster
Duplicated X-chromosome,
half meaningless constitutive
heterochromatin

Obvious species-specific changes which one occasionally sees in the X or the Y of exceptional mammalian species, are functionally meaningless changes, since they are due either to the addition of variable chunks of dispensable constitutive heterochromatin to the X or the Y, or the terminal fusion between the X and an autosome (the so-called XX/XY_1Y_2 system), or the Y and an autosome (the so-called $X_1X_1X_2X_2/X_1X_2Y$ system). Neither type of terminal fusion involving an autosome alters the chromosomal sex-determining mechanism in the slightest. The X of the XX/XY_1Y_2 system, in reality, is (X+A), while Y_2 is nothing but an intact homologous autosome, A (Fig. 1-7). On the other hand, only X_1 of $X_1X_1X_2X_2/X_1X_2Y$ system is the real X, while the Y in reality is (Y+A) and X_2 is nothing but an intact homologous autosome, A (Fig. 1-8).

The principle of evolutionary conservation also applies to the nuclear-cytosol androgen-receptor protein apparently specified by the X-linked *Tfm* locus. No matter whether this regulatory protein comes from man, rats or mice, it always shows the same subunit size made up of about 500 amino acid residues, and nearly identical relative binding affinities to 5α-dihydrotestosterone and testosterone.

Although rigid evolutionary conservation prevails everywhere within the functionally critical parts of the genomic DNA, such observed molecular conservatism should not be construed to imply that evolutionary divergences, which one can see with the naked eye, are mere illusions, for man is clearly different from chimps, and a camel is decisively not a horse. Rather, this apparent conflict between molecular conservatism and morphological diversity reveals to us the extreme subtlety of evolutionarily meaningful genetic changes. Evolution does not pro-

Fig. 1-6. The evolutionary conservation of mammalian X-chromosomes is also evident in their banding patterns visualized by a modern Giemsa banding technique. It should be noted that all the mammalian X-chromosomes possess two major bands, one thicker than the other. (After S. Pathak and A. D. Stock. *Genetics* **78**, 703 (1974).)

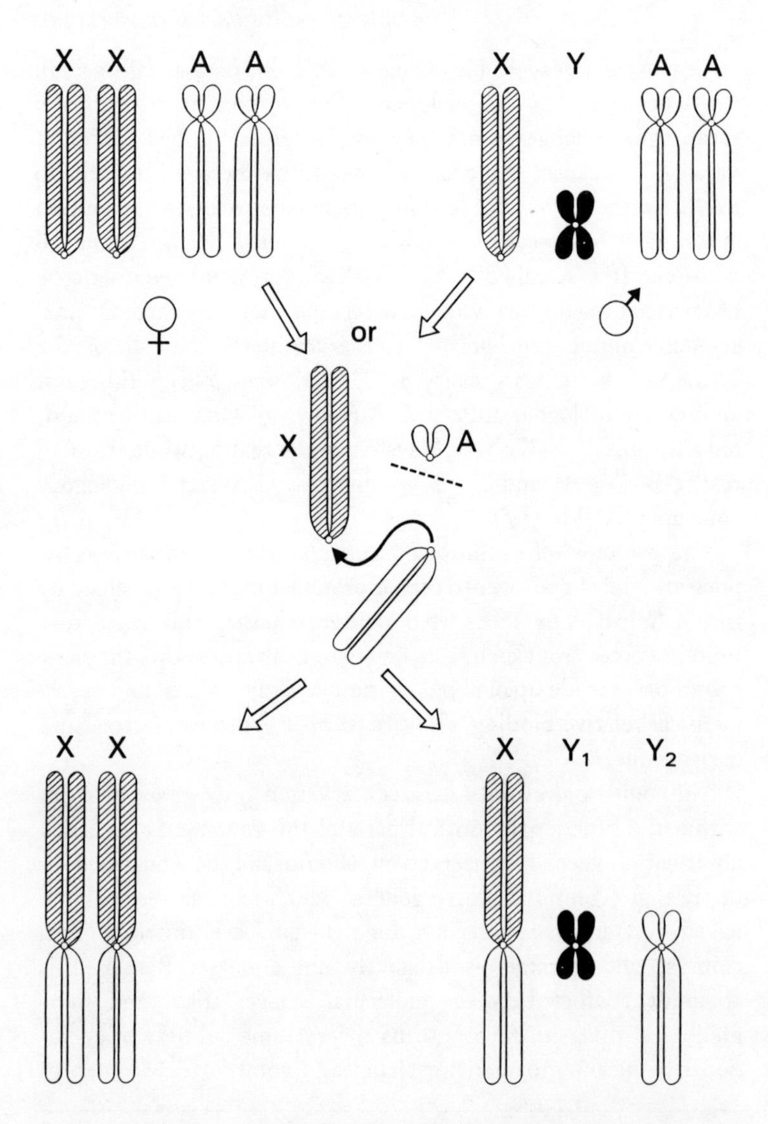

Fig. 1-7. A change from an ordinary XX/XY system (top) to the so-called XX/XY_1Y_2 system (bottom) by a simple terminal fusion between the original X and an autosome (middle). The original X is hatched, the original Y is drawn solid black, an original pair of autosomes is drawn in outline. (After K. Fredga. *Philosophical Transactions of the Royal Society of London, Series B* **259**, 15 (1970).)

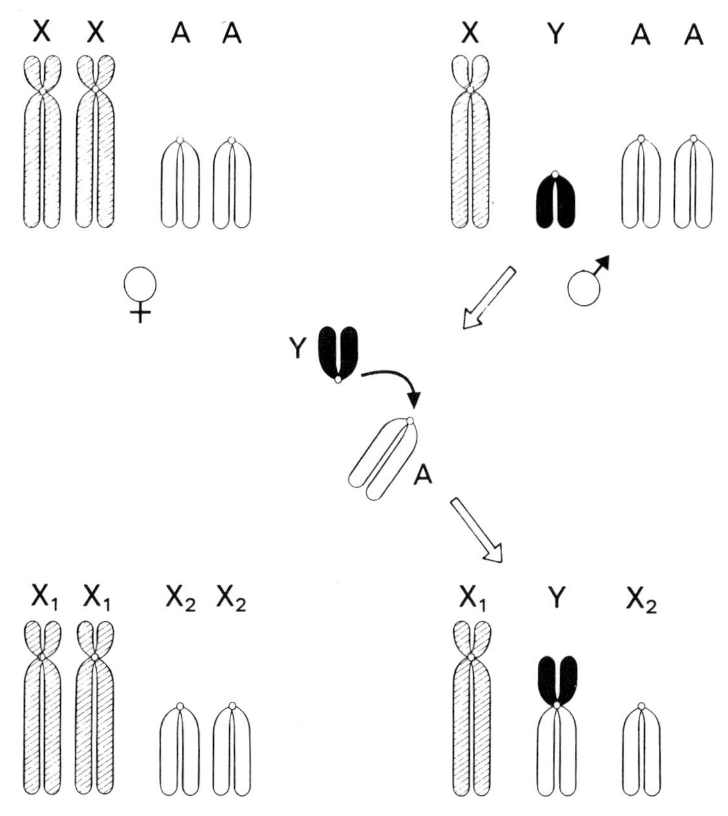

Fig. 1-8. A change from an ordinary XX/XY system (top) to the so-called $X_1X_1,X_2X_2/X_1X_2Y$ system (bottom) by a simple terminal fusion between the Y and an autosome. The original X, Y and a pair of autosomes are marked in the same way as in Fig. 1-7. (After K. Fredga. *Philosophical Transactions of the Royal Society of London, Series B* **259**, 15 (1970).)

ceed by taking such a gross step as changing the active site amino acid sequence for essential household enzymes.

If man could communicate with other mammals they would be likely to tell us that we are the oddest creature on this land, being stark naked. Yet such a distinguishing characteristic of *Homo sapiens* has apparently been obtained by a single recessive mutation, for the recurrence of a revertant dominant mutation, which would restore man's fur coat, has been well recorded in history.

Fig. 1-9. The human female is the only primate in which significant mammary development occurs *before* the first pregnancy. The stromal cells of the human breast have become oestrogen-sensitive, as the breast has developed a new role as sex attractant.

Clearly, this locus does not specify anything as obvious as a hair keratin protein, for even naked apes retain luxuriant hair growth in certain parts of their body, public as well as private.

Although the meaningful genetic content of the X as well as the Y, and hence the basic sex-determining mechanism of mammals, has remained inviolate throughout the age of mammals, extremely subtle shifts in peripheral areas of the sex-determining mechanism have, no doubt, contributed greatly to their tremendous adaptive radiation. Evolutionary changes in sex-related,

Fig. 1-10. In chimpanzees, there is no obvious sexual dimorphism with regard to their facial hair growth. In man, the hair growth around the lips (beard and moustache) became androgen-dependent, thus accentuating the degree of sexual dimorphism.

hormone-dependent functions are particularly significant, for they tend to cause drastic alterations in individual behaviour and the social organization of a species.

Aside from being stark naked, two other characteristics set the human female apart from her closest relative, the chimpanzee. A human female is almost always 'in heat' regardless of the stage of the menstrual cycle, and her breasts are always prominent even when she is not lactating; this is not due to the constitutive growth of mammary tissue, but to the oestrogen-dependent growth of fatty tissue (Fig. 1-9). Similarly, a human male advertises his sexual maturity by growing luxuriant hair around his mouth, while a chimpanzee male has no such indicator (Fig. 1-10). It is obvious that, inside the female human brain, the control of sexual behaviour, which was originally cyclic and dependent on female hormones, has now become constitutive. Similarly inside the human breast, fat cells, which were originally indifferent to oestrogen, have now become oestrogen-dependent. In a similar manner, hair follicles in the skin region around the human mouth must have suddenly become androgen-dependent.

The development of sexual reproduction

A very prolonged infancy characterizes human development, and this characteristic apparently served as one of the corner stones for the development of human intelligence, for a long infancy is conducive to learning. Infants can afford to have a long learning period only if their parents develop a lasting pair bond. Obviously, such a bond has to be based upon an extraordinarily strong and persistent sexual attraction between male and female. There is little wonder that human females and males continually advertise their sexuality to one another in order to cement the lasting bond. The point is that such conspicuous and very significant evolutionary changes have been brought about by subtle sleights of hand without having to alter kinetic properties of the existing steroid hormone receptor proteins. By the very simple act of switching on a repressed steroid hormone receptor locus, a cell type that was formerly beyond the reach of hormonal influence now joins the ranks of the target cells for that hormone. In such manner extremely subtle evolutionary changes in the peripheral areas of the chromosomal sex-determining mechanism have contributed greatly to the adaptive radiation of mammals, and the eventual emergence of man.

SUGGESTED FURTHER READING

A discussion on determination of sex. Organized by G. W. Harris and R. G. Edwards. *Philosophical Transactions of the Royal Society of London, Series B* **259**, 1 (1970).

Sex-reversed mice: XX and XO males. B. M. Cattanach, C. E. Pollard and S. G. Hawkes. *Cytogenetics* **10**, 318 (1971).

Normal spermatozoa from androgen-resistant germ cells of chimaeric mice and the role of androgen in spermatogenesis. M. F. Lyon, P. H. Glenister and M. L. Lamoreaux. *Nature, London* **258**, 620 (1975).

Evolutionary conservation of H-Y ('male') antigen. S. S. Wachtel, G. C. Koo and E. A. Boyse. *Nature, London* **254**, 270 (1975).

Possible role for H-Y antigen in the primary determination of sex. S. S. Wachtel, S. Ohno, G. C. Koo and E. A. Boyse. *Nature, London* **257**, 235 (1975).

Ancient linkage groups and frozen accidents. S. Ohno. *Nature, London* **244**, 259 (1973).

Major regulatory genes for mammalian sexual development. S. Ohno. *Cell* **7**, (1976).
Intersexuality in the Animal Kingdom. R. Reinboth. Berlin; Springer-Verlag (1975).
Protochordata, Cyclostomata and Pisces. S. Ohno. Berlin and Stuttgart; Gebrüder Bornträger (1974).

NOTE ADDED IN PROOF

A fascinating mechanism has just been discovered in the wood lemming *Myopus schisticolor* of Northern Europe and Asia which results in an unequal sex ratio at birth with a considerable excess of females. There are apparently two classes of normal, fertile females in the population, one with the usual XX karyotype, and one with an XY karyotype indistinguishable from that of normal males. However, the XY females are H-Y antigen negative while the XY males are positive. The XY females have XX germ cells, which have presumably arisen by mitotic non-disjunction and selective loss of the Y-chromosome. The authors postulate that there must be an X-linked gene that represses the male-determining effects of the Y-chromosome; thus if an XX female heterozygous for this gene were to mate with a normal male, there would be a 1 : 3 ♂ : ♀ sex ratio in the offspring; if an XY female were to mate with a normal male, all the offspring would be female.

(Fertile XX- and XY-type females in the wood lemming *Myopus schisticolor*. K. Fredga, A. Gropp, H. Winking and F. Frank. *Nature, London* **261**, 225 (1976).)

2 Evolution of viviparity in mammals

G. B. Sharman

All life began in the sea, and the earliest sexually reproducing, multicelled animals shed their germ cells into the surrounding water where fertilization occurred and embryos developed. In *oviparous* animals, like these with *external fertilization*, the chance of union between male and female germ cells is greater if the two sexes are in close proximity at spawning time. Methods for bringing the sexes together were adopted independently by different oviparous animals, and include the release of chemical attractants by one or other sex and the use of display behaviour to attract a partner.

A better way of ensuring union of gametes is to have *internal fertilization*, and this is generally done by introducing spermatozoa into the female reproductive tract at about the time when the eggs are ripe. With externally fertilized eggs problems arise over the need to have protective coats around the eggs whilst still permitting spermatozoa to enter; on the other hand, eggs fertilized internally can have coats added after sperm penetration. An obvious further step in the evolution of parental care is to retain the egg in the female reproductive tract for at least part of embryonic development. In *ovoviviparous* animals, birth and egg hatching are more or less simultaneous events – the egg with its yolk, albumin and shell is retained in the reproductive tract until embryonic development is complete. In truly *viviparous* animals, yolk, albumin and egg shell are reduced or absent, and the developing embryo becomes parasitic on the mother. The female tract, primitively a simple tube or duct for passing the unfertilized egg to the exterior, now becomes an organ of gestation and is profoundly modified.

The most familiar kind of viviparity – that exhibited by eutherian (placental) mammals – was accompanied during its evolution by the development of reproductive processes essential for prolonging the stay of the embryo in the uterus; many of these mechanisms have been discussed in other books in this series. The ruptured ovarian follicle from which the egg was shed is transformed into an endocrine gland, the corpus luteum (Book 1, Chapter 2), which secretes progesterone, the hormone of pregnancy (Book 3, Chapters 1 and 3). Embryos must travel down the oviduct and arrive in the uterus at the right time, and, once there, become spaced and oriented for implantation; trophoblast–endometrial interactions then become enormously important (Book 2, Chapter 1). The maintenance of pregnancy depends on complex hormonal controls (Book 3, Chapter 4) and on adaptations that allow fetal survival in what could be an immunologically hostile uterus (Book 4, Chapter 4); the same hormones are sometimes used in different ways by different animals (Book 4, Chapter 1). Viviparity is terminated by another complex sequence of endocrine changes that result in the birth of the fetus (Book 2, Chapter 3).

Evolution can be studied in several ways. One involves a survey of the various evolutionary adaptations that can be seen in bony structures preserved in fossils. Soft tissues are seldom preserved and so the fossil record can provide us with at best indirect evidence about the evolution of viviparity. Most of the surviving animals with long fossil records are oviparous, whereas the marsupials and placental mammals, which have the shortest fossil record amongst the vertebrates (Fig. 2-1), are all viviparous. Fortunately, the reproductive organs of viviparous animals are sufficiently like those of oviparous species for us to be certain of their homology since, as Milne Edwards observed, 'Nature is prodigal in variety but niggard in innovation'. Egg transport ducts have been modified as organs of gestation, egg coats as structures for intrauterine nutrition and excretion, and the gonadal steroids of oviparous animals as the pregnancy hormones of viviparous species. To understand the ways in which viviparous reproduction has evolved, we should therefore begin with oviparous forms, and try to determine the adaptations that have

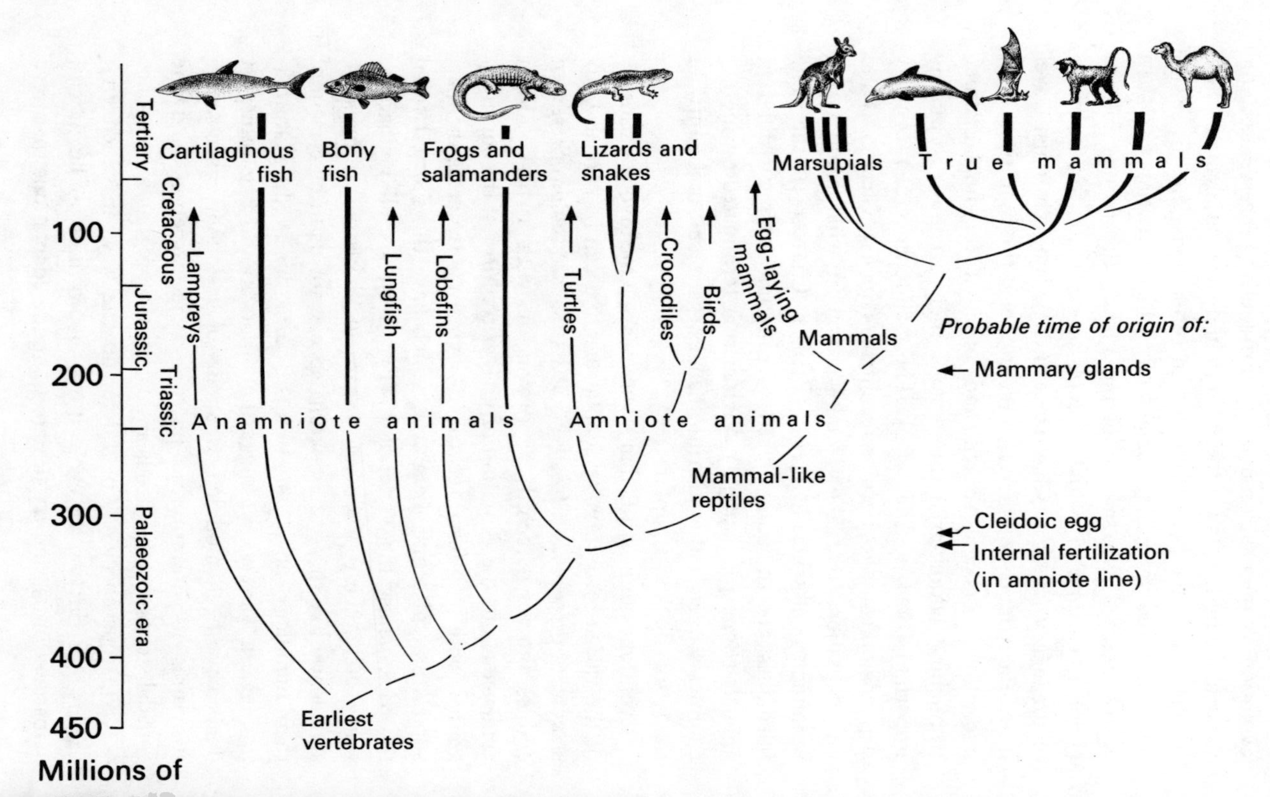

Tertiary
Cretaceous
Jurassic
Triassic
Palaeozoic era
100
200
300
400
450
Millions of
Cartilaginous fish
Bony fish
Frogs and salamanders
Lizards and snakes
Marsupials
True mammals
Lampreys
Lungfish
Lobefins
Turtles
Crocodiles
Birds
Egg-laying mammals
Mammals
Anamniote animals
Amniote animals
Mammal-like reptiles
Earliest vertebrates
Probable time of origin of:
Mammary glands
Cleidoic egg
Internal fertilization (in amniote line)

made viviparity a possible and indeed highly successful mode of reproduction.

INTERNAL FERTILIZATION – A PREREQUISITE OF VIVIPARITY

Contemporary jawless vertebrate animals (Agnatha), such as the lampreys, have no specialized ducts for transporting the germ cells from the gonad to the exterior of the body. Eggs and spermatozoa are shed directly into the body cavity, and make their way through pores in the body wall to the surrounding fresh water where external fertilization takes place. Prior to spawning a 'nest' is built in the stones and gravel of the river bed and the male wraps his body around the female, so that extruded eggs and spermatozoa are released near together and fertilized eggs are deposited in the nest.

In cartilaginous fishes (Chondrichthyes) and amphibians, the kidney (archinephric) ducts became modified for the transport of male germ cells to the exterior of the body (Fig. 2-2). Thus the first step in the evolution of a sperm transport system in the male made use of a duct that was already being used in another body system. In the female, a new structure, the Müllerian or paramesonephric duct, sometimes simply called the oviduct, was developed for egg transport (Fig. 2-2).

Unlike the cartilaginous fishes and amphibians, and also the reptiles, birds and mammals, the sex ducts of bony fishes (Osteichthyes) are homologous structures in the two sexes. The testis cavities and ovisacs develop in the embryo as leaf-like extensions of the gonads, and these are extended posteriorly into tubes which follow the mesenteries of the gonadal ridges and empty by pores at the posterior end of the body. This is an

---

Fig. 2-1. Generalized diagram of the evolution of vertebrate animals. Evolutionary lines with *some* (or all) viviparous species that persist to the present are shown thick; lines in which viviparity was not developed are shown thin and are not carried to the top of the diagram. One or a few of the many viviparous forms in each animal class are shown at the top.

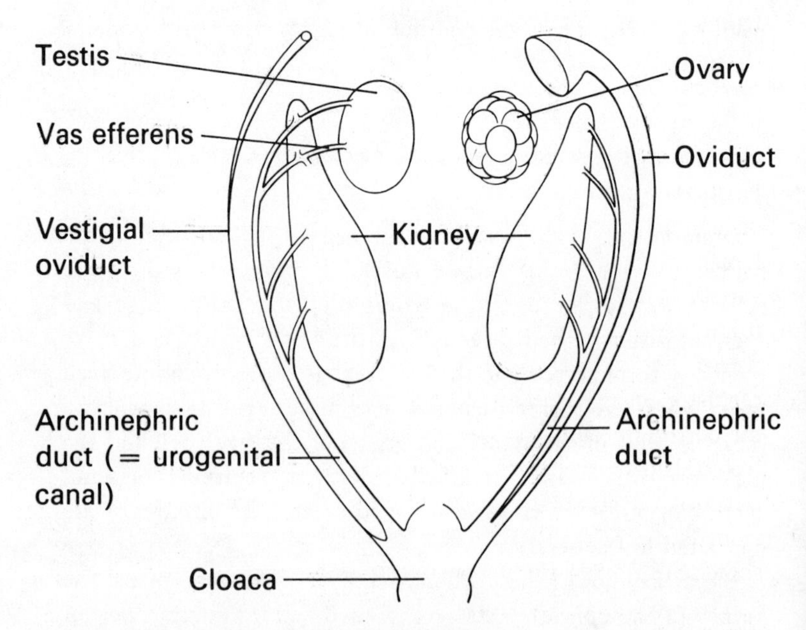

Fig. 2-2. Generalized drawings of left half of the male and right half of the female excretory and reproductive systems (urogenital system) in cartilaginous fishes and amphibians. The excretory duct of the male is modified for sperm transport to the exterior of the body, whereas the oviduct of the female is an evolutionary innovation.

example of convergent evolution; the gametic ducts of bony and cartilaginous fishes are derived from different structures, but have come to serve the same purpose.

The advent of internal fertilization

Many fishes and amphibians have simple egg and sperm ducts which help to prevent the germ cells getting lost in the body cavities, but fertilization still takes place in water, outside the body. A more efficient system, which apparently arose independently in several animal groups, permits fertilization of the eggs before they leave the oviduct.

Some salamanders use the cloaca to take up spermatophores

(packets of spermatozoa) which have been deposited by males on twigs or stones, thus achieving internal fertilization without physical contact between the two sexes. In other amphibians and most birds spermatozoa are passed from the male to the female cloaca during courtship. The efficiency of internal fertilization has been improved during evolution chiefly by the development of copulatory organs which deposit spermatozoa within the female reproductive tract. Male copulatory organs exist in a variety of forms, indicating numerous parallel lines of evolution for internal fertilization. The modified medial borders of the paired pelvic fins (the claspers) serve this function in male cartilaginous fishes, while in many bony fish such as the teleosts the copulatory organ is an unpaired midline structure developed from the anal fin. Male snakes and most male lizards have not one but a pair of copulatory organs (hemipenes), which are formed from lateral cloacal wall tissue, whereas turtles, crocodiles and some lizards have an unpaired intracloacal penis.

The advent of internal fertilization also meant that the spermatozoa had much less need to swim to find eggs to fertilize – in mammals, at least, ejaculation and the movements of oviduct and uterine walls are more important in sperm transport (Book 1, Chapter 5). Spermatozoa deposited in the female tract also find themselves in a better physiological environment than do those shed into the water, which is usually of a different ionic concentration to that of the germ cell cytoplasm. Finally, it is worth remembering that only those animals that had already acquired internal fertilization were able to achieve complete emancipation from a wet environment and colonize the dry land.

OVIDUCTS OF MAMMALS

Mammals are derived from mammal-like, or therapsid, reptiles (Fig. 2-1) which are known only from 200-million-year-old fossils. What method of reproduction the therapsids used is of course unknown, but since they lived on the land they must have developed internal fertilization, and we suspect that they were oviparous. Even before they gave rise to the mammals, the therapsids already had a long history of separate evolution from the remaining reptiles, which may have extended back almost to the time of the amphibian grade of organization. For this reason, comparing the reproductive systems of the most primitive living mammals, the egg-laying monotremes, with those of modern reptiles is unwise because monotremes and living reptiles have followed separate evolutionary paths since Carboniferous times.

We have already discussed the way in which the sperm ducts of mammals, birds, reptiles and cartilaginous fishes are derived from the archinephric ducts of the embryo; but the higher vertebrates develop another excretory duct – the metanephric duct or ureter – in both sexes during later embryonic life. The female reproductive tracts of the higher vertebrates are derived from the embryonic Müllerian or paramesonephric ducts, again as happens in cartilaginous fishes and amphibians. This duct began to become specialized in oviparous animals, when one or more regions were developed for the deposition of substances, such as the shell, round the fertilized egg.

The term 'oviduct' is commonly restricted to the Fallopian tube in mammals, but since we are concerned here with the evolution of the entire length of the female reproductive tract, we will use 'oviduct' to describe all the Müllerian duct derivatives.

The monotreme double oviduct

There are three genera of living monotremes, each containing only a single species. The reproductive systems of the female platypus (*Ornithorhynchus*) and the two echidnas (*Tachyglossus* and *Zaglossus*) (Fig. 2-3) consist of two completely unfused oviducts and these open separately into a urogenital sinus, which also receives the opening of the bladder. Each oviduct is divisible into an anterior tubal portion and a slightly enlarged, more posterior region, termed the uterus; there is no true vagina as in other mammals. The ureters do not enter the base of the bladder but open into the urogenital sinus, so that excretory products

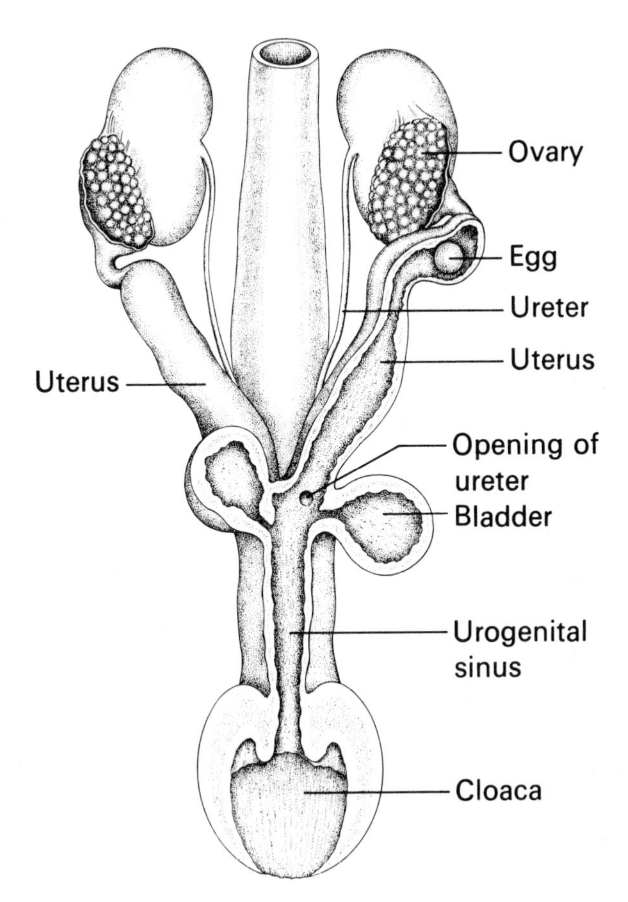

Fig. 2-3. Female urogenital system of the echidna. (Based on drawings by F. Wood Jones, *The Mammals of South Australia*, Part I, fig. 14. Adelaide; Government Printer (1923); and Mervyn Griffiths, *Echidnas*, fig. 53. Oxford; Pergamon Press (1968).)

from the kidneys must cross the sinus prior to storage in the bladder (see Fig. 2-3). The urogenital sinus terminates in a cloaca which also receives the posterior opening of the alimentary canal, so that a combined rectal, urinary and genital opening is found. The monotremes are the only mammals that have a true cloaca and they are identified by this feature – their name comes from the Greek *mono-* meaning single, and *-trema*, a hole, whereas the remaining mammals are sometimes grouped as Ditremata because rectal and urogenital openings are more or less separate.

In the upper (tubal) portion of the monotreme oviduct a thin layer of what was originally called 'albumen' is deposited around the egg, while in the more distal glandular part the basal layer of shell is added to the existing egg coats. Leon Hughes has shown by histochemical methods that the 'albumen' layer of the monotreme egg is actually a mucoid coat composed of acidic glycoprotein, and that the egg shell is indeed keratinous in structure. Two further layers of shell, the rodlet and outer layers, are deposited around the egg by secretions of uterine glands. Because of their shell-secreting functions, parts of the monotreme oviduct may be likened to the shell gland of oviparous animals that have internal fertilization; but the uterine portion of the oviduct has a further function. The same glands that produce the rodlet layer of shell also produce embryotrophe, often called 'uterine milk', which is absorbed by the developing egg as it passes down the oviduct. Perhaps the uterine glands of monotremes (and other mammals) are a legacy from the shell-producing glands of oviparous ancestors which have evolved the additional function of supplying nutrient substances to the developing intrauterine embryo. In any event the oviparous monotremes exhibit the earliest known grade of mammalian viviparity – nutrition and growth of the embryo, surrounded by its shell membranes, within the oviduct.

The marsupial compromise oviduct

The two uteri of marsupials are unfused, as in the monotremes, but open into a vaginal region rather than into the urogenital sinus, and the ureters now enter the base of the bladder. The female marsupial reproductive tract (Book 4, Chapter 1, Fig. 1-3) appears complicated when compared to that of eutherian mammals, but the differences are solely due to the position of the ureters with respect to the oviducts. There are two lateral vaginae which loop to each side of the reproductive system like two handles of a vase; posteriorly they open into a urogenital sinus, as do the oviducts of the monotremes. The vaginae cannot fuse into the single midline structure of the eutherian mammals, because the ureters lie between them. These lateral vaginae are the ancestral egg ducts of marsupials, but whether they once functioned also as birth canals is unknown. If they did, they would have precluded the birth of large young, since they are tortuous and narrow. They certainly transport the spermatozoa up to the uterus and Fallopian tubes, where the eggs are fertilized (Book 4, Chapter 1), but the young are born through a midline birth canal which connects the two uterine openings directly to the urogenital sinus. Thus, in contrast to other viviparous animals, the ascending spermatozoa and the descending full-term fetus follow different paths.

The marsupial reproductive system is unique, but shows considerable variation between species (Fig. 2-4). In the pouchless marsupial 'mouse' *Antechinus*, the septum separating the vaginal culs-de-sac is present even in females that have given birth and the Müllerian ducts persist as separate canals right down to the urogenital sinus. The pseudovaginal canals open at each parturition and close again as soon as birth occurs. In the native cat *Dasyurus*, which gives birth to a large number of young each weighing a mere 12 mg, an essentially similar condition prevails. A short pseudovaginal canal which opens anew at each parturition is found in the brush possum *Trichosurus*, which gives birth to a single young weighing 200 mg, while the larger kangaroos, *Macropus* spp., which produce 'giant' single young weighing nearly a gramme have a patent midline vagina following the first parturition. Thus the kangaroos have achieved the equivalent of the eutherian midline vagina, but by rather a different method.

Evolution of the eutherian fused oviduct

The reproductive tract of female eutherian mammals is unique in having a fused midline structure, the vagina, which lies *between* the two ureters. This was not possible in the marsupials, because

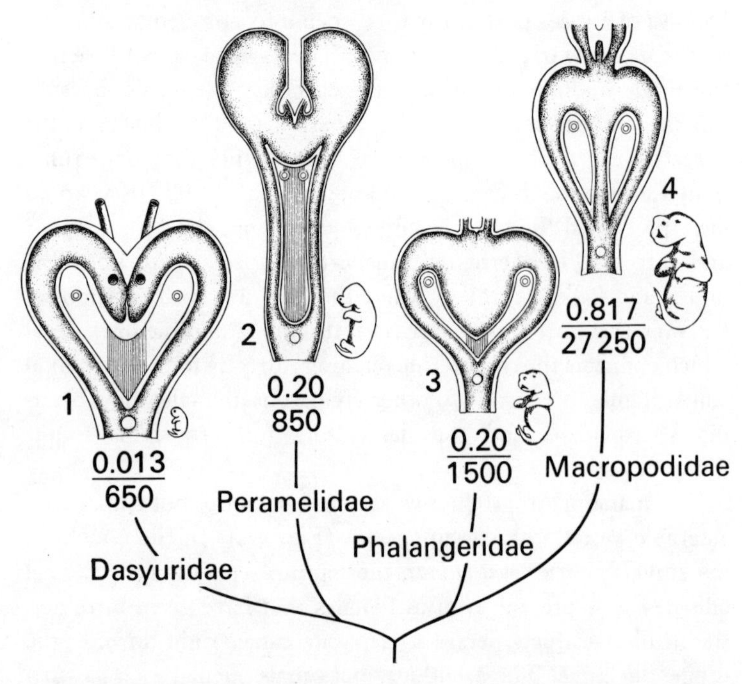

Fig. 2-4. End-products of evolution of the marsupial vaginal apparatus (not drawn to scale) as seen in representatives of four marsupial families. 1, Eastern native cat *Dasyurus viverrinus*; 2, long-nosed bandicoot *Perameles nasuta*; 3, brush possum *Trichosurus vulpecula*; 4, red kangaroo *Macropus rufus*. Newborn young are drawn to scale; the fractions represent the weight of young at birth/weight of the mother (in grammes). Double circles show the positions of ureters and open circles the position of the opening of the bladder into the urogenital sinus. Vertical hatching indicates tissue that opens to form the pseudovaginal canal at parturition (the canal is permanently open in kangaroos (Macropodidae) after the first parturition; see also Fig. 1-3 in Book 4 of this series). The native cat, with median vaginal septum retained in parous females, has a vaginal apparatus most like that of the hypothetical ancestral marsupial. (Based on G. B. Sharman, 'Marsupials and the evolution of viviparity', fig. 1-5. In *Viewpoints in Biology*, Vol. 4. Ed. J. D. Carthy and C. L. Duddington. Butterworth; London (1965).)

it would have occluded the ureters (Fig. 2-5). Both Wolffian and Müllerian ducts are present in the early eutherian embryo but one duct system is eliminated from each sex as sexual differentiation

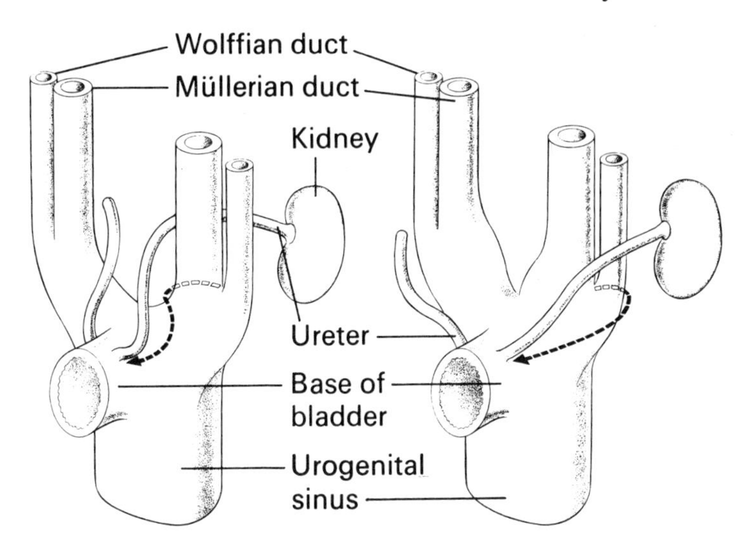

Fig. 2-5. The arrangement of excretory and sex ducts at an early stage of development in true mammals compared to that in marsupials. In true mammals (*right*) the ureters grow forwards and pass *outside* the Wolffian and Müllerian ducts whereas in marsupials (and in all other amniote animals) the ureters grow forwards *between* the Wolffian and Müllerian ducts. The direction of longitudinal growth and the movement of the points of origin of the ureters are shown by arrows. The remainder of one ureter and the attached kidney have been omitted in each diagram. In both groups the ureters originate as buds on the dorsomedial side of the Wolffian duct and after shifting in different directions they open into the base of the bladder primordium.

proceeds (Book 2, Chapter 2). Fusion of the embryologically paired oviducts does not end in the vaginal region; the uterine portions also show varying degrees of fusion, culminating in the completely fused uterus or uterus simplex of primates; abnormalities of the human uterus reveal its bilateral origins (Fig. 2–6). One of the important advances in mammalian viviparity was the development of a reproductive tract that allowed the gestation and birth of large young; in this respect the eutherians had a considerable advantage over the marsupials.

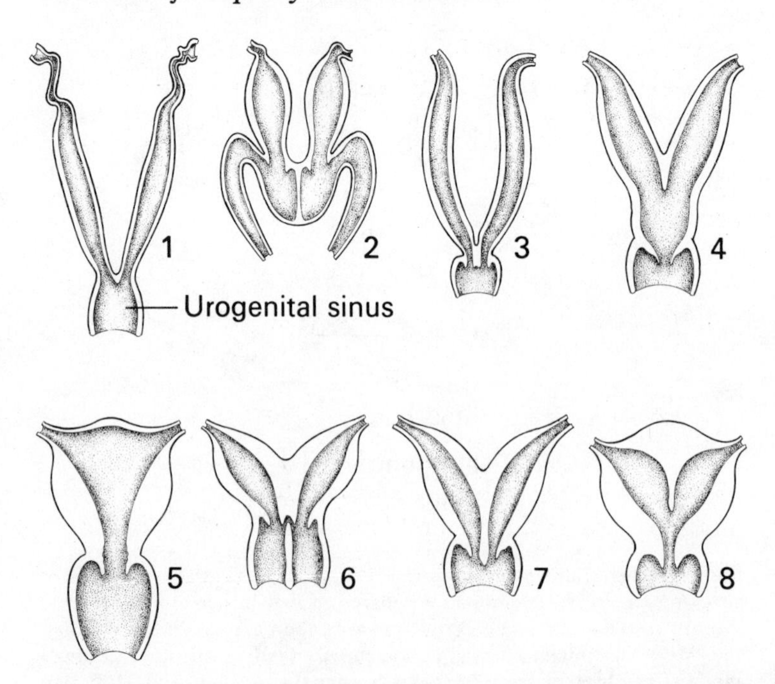

Fig. 2-6. The mammalian uterus. 1, Monotreme condition; the two uteri open separately into the urogenital sinus which also receives the openings of bladder and ureters; there is no vagina. 2, Marsupial condition; separate uterine openings into vaginal culs-de-sac which connect with paired lateral vaginae. 3, Duplex uterus as found in rabbits; separate cervical openings into single vagina. 4, Bicornuate uterus, found in most domestic and laboratory mammals. 5, Simplex uterus of most primates. 6–8, Anomalies of human uterus: 6, double uterus and vagina caused by failure of Müllerian ducts to fuse; 7, double uterus; 8, subseptate uterus.

ENCLOSED EGGS AND EMBRYONIC MEMBRANES

A primary distinction is made between the two major groups of animals that are adapted to live on land. Frogs, salamanders and other amphibians are classed as Anamniota and reptiles, birds and mammals as Amniota, the names deriving from the absence or presence of the amniotic membrane found in the embryos of all members of the second group. The amnion first arose in the egg of oviparous and presumably aquatic animals, that laid their eggs on the land. Laying eggs in the water in the days when all vertebrates were aquatic was presumably an open invitation for them to be eaten, so laying eggs on land not only gave them a better chance of completing development, but was also the first step in the evolution of mammalian viviparity. This new type of egg is called the amniote or cleidoic ('enclosed') egg; the embryos contained therein are independent of their surroundings, apart from respiratory requirements.

The steps by which the cleidoic egg may have evolved from the egg of anamniotes are shown in Fig. 2-7. The amniotic cavity and extraembryonic coelom are filled with fluid which takes the place of the water required for successful anamniote development. The allantois, a diverticulum of the gut arising behind the opening of the yolk-sac, is homologous with the urinary bladder, a structure that makes its first evolutionary appearance in amphibians. In oviparous amniotes the allantois functions as a storage bag in which the excretory products of the embryo are placed until hatching.

Some fish and the tadpoles of frogs excrete much nitrogen as urea; this is a toxic substance, but also highly soluble and the large volume of water surrounding the animal ensures its rapid diffusion. For the embryo in the cleidoic egg to produce urea would be disastrous as this could not escape from the closed system. The evolution of the cleidoic egg was therefore dependent on the simultaneous or prior evolution of the ability to excrete nitrogenous wastes in the form of insoluble uric acid, which is stored in the allantois.

In all oviparous vertebrates the allantois, with its plentiful supply of blood vessels, eventually makes contact with the chorion (Fig. 2-7), the layer of embryonic tissue just inside the egg shell. The egg shell serves to prevent desiccation but is sufficiently porous to allow gaseous exchange, and the allantois acts not only as a dustbin but also as the embryonic 'lung' (see Fig. 2-7). The amnion, allantois and chorion are collectively called the extraembryonic membranes because they do not contribute to the embryo proper and are discarded at hatching.

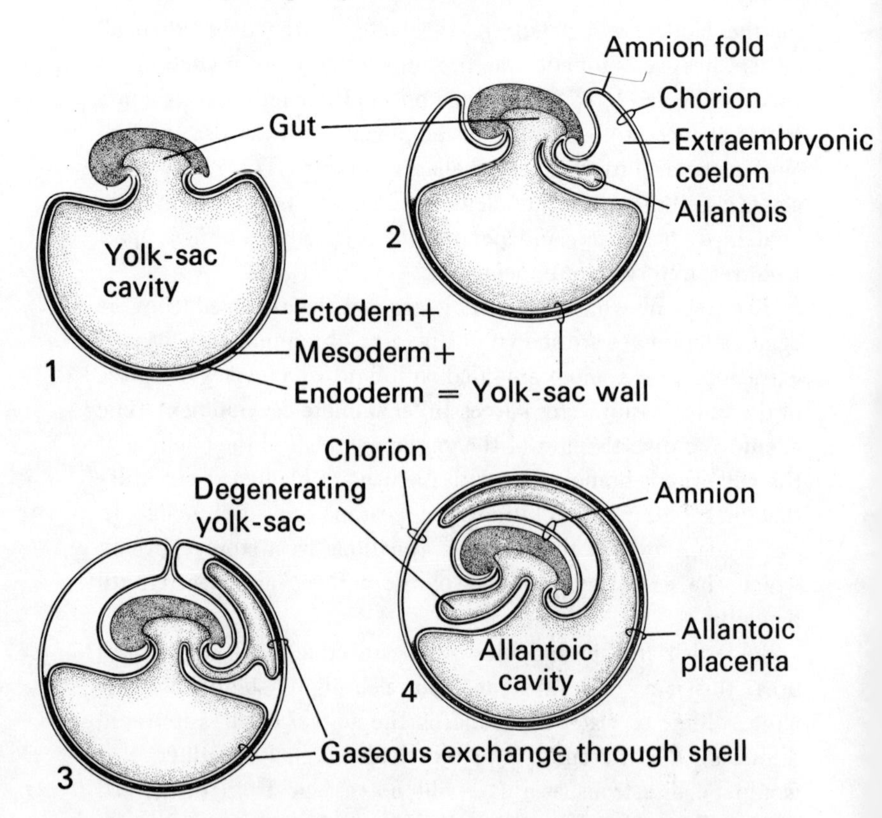

Fig. 2-7. Longitudinal sections through the developing embryos of anamniote and amniote animals. 1, Anamniote condition; there are no extraembryonic membranes and the yolk-sac is eventually incorporated into the gut of the young animal. 2, Hypothetical intermediate stage between anamniote and amniote animals: formation of amnion folds and splitting of mesoderm (thick line) into two layers enclosing extraembryonic coelom. 3, Condition found in all oviparous amniotes before complete fusion of amnion folds above the embryo; the egg shell (not drawn) surrounds the embryo and its membranes. 4, Condition found in all true mammals; the yolk-sac is regressing, the egg shell is abolished and exchange between uterus and embryo is by means of the allantoic placenta.

Enclosed eggs and embryonic membranes

Many oviparous animals abandon their cleidoic eggs, buried or camouflaged, and allow embryonic development to proceed without parental intervention. Some of them may guard the eggs until they hatch and then assume parental care of the young, while others, such as the mound-building birds of the Australian region, make an elaborate chamber of fermenting vegetation in which the eggs are incubated. The evolution of homoeothermy, or the ability to control body temperature independently of the environment, led to the adoption of egg incubation by the warmth of the body of one or other parent, a method still used by the egg-laying monotremes.

Since the maximum degree of protection is given by retaining the eggs in the oviduct during embryonic development, it is not surprising that true viviparity has supplanted oviparity in all mammals, except the monotremes, and in many reptiles. During gestation in the oviduct the extraembryonic membranes and yolk-sac (Fig. 2-7) have become further modified as organs of nutrient and waste-product exchange between maternal organism and developing young.

We have now considered three of the major evolutionary adaptations that enabled viviparous animals to retain developing embryos in their reproductive tracts; internal fertilization, the specialization of the female reproductive organs, and the development of the extraembryonic membranes to allow nutrient and waste-product exchange between the embryo and its environment. These and many other innovations are necessary for successful mammalian-type viviparous reproduction. However, many vertebrates with internal fertilization, specialized reproductive systems and embryonic membranes have opted to retain oviparity, and it would be wrong to regard them as hypothetical ancestors of viviparous animals. Similarly, many animals that do not have amniote-type extraembryonic membranes and may even lack internal fertilization have become viviparous. Some of these unusual types of viviparity deserve detailed discussion.

ADAPTATIONS FOR VIVIPARITY IN MONOTREME AND MARSUPIAL MAMMALS

Female echidnas (spiny anteaters) are known to have laid fertile eggs 17 and 18 days after being separated from a male and, in one instance, a fertilized egg was found in the uterus of a female that had been separated from the male for 34 days. Mervyn Griffiths considers that viable spermatozoa may be stored in the female tract for many days after mating, so at the moment we cannot calculate the exact period of intrauterine gestation in monotremes. The embryo in the newly laid egg is hardly more developed than a chick on the second day of incubation. The single egg (rarely two, or even three) is incubated after laying in a brood pouch on the ventral surface of the female's body for a further 10 days before hatching from the shell (Fig. 2-8). The newly hatched young is then reared in the pouch close to the paired mammary glands, both of which are functional. Not surprisingly, the young animal vacates or is ejected from the pouch when it begins to develop the spiny covering that characterizes the adult, and is then maintained in a nest for a period of several weeks.

The platypus lays one, two or three eggs, 'twins' and 'triplets' being cemented together. The eggs are incubated for an unknown period, apparently clasped between the female's flattened tail and her abdomen while she occupies a plugged burrow.

There are some 230 living species of marsupials and we know something about reproductive processes of only a score or so. Some species keep their young in the uterus for an even shorter time than the monotremes, while others have longer gestation periods than some eutherians. Pouch life of the young may be as short as 7 weeks, as in some bandicoots, or it may last almost a year as in the grey kangaroo. Generally speaking, the marsupial young on leaving the pouch is at an equivalent stage of development to a newborn eutherian, and a further period of maternal milk feeding, analogous to the suckling period of placental mammals, occurs after the young leaves the pouch.

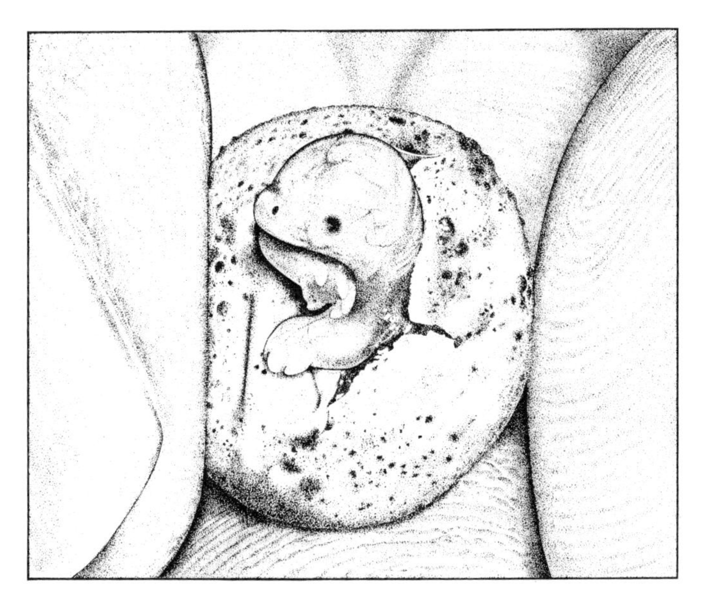

Fig. 2-8. Young of the echidna, an egg-laying mammal, in the process of breaking out of the egg shell. Hatching occurs in a brood pouch on the ventral surface of the female parent's body (After Mervyn Griffiths. 'The life of the echidna', *Australian Natural History* **17**, 222–6 (1972).)

Lactation in monotremes and marsupials

Lactation is one of the few major features of reproduction common to monotremes, marsupials and eutherians. Jim Linzell pointed out that the mammary gland has in effect yielded to the placenta part of the function of feeding the fetus during the course of mammalian evolution. Newly hatched monotremes such as the echidna (Fig. 5-1*a*, Book 3) weigh less than half a gramme and could not be raised without milk feeding, whereas eutherians give birth to comparatively robust young which, in some species, can be raised artificially without milk. Thus, phylogenetically speaking, lactation preceded viviparity and arose as an adaptation for the nourishment of young of oviparous species hatched from the egg at early developmental stages.

In the non-breeding season the lobules of echidna mammary

glands consist of solid cords of cells, but Mervyn Griffiths found that, during the breeding season, some mammary glands become tubular and may even secrete milk, in the absence of an egg or newly hatched young. Perhaps the mammary glands can develop in response to cyclic ovarian changes in the absence of a pregnancy.

There is no doubt about the significance of hormones from the corpus luteum for mammary gland development and the initiation of lactation in marsupials. The development of the mammary glands in pregnant and non-pregnant females is identical, and newborn young can be fostered onto the teats of virgin females in a variety of marsupial species, where they will initiate lactation and grow normally. Thus the initiation and maintenance of lactation in marsupials is hormonally dependent only upon an active corpus luteum. The marsupial situation is thus comparable to that of the monotremes, which suckle their young but can hardly be said to experience pregnancy in the mammalian sense since they lay eggs.

The corpus luteum

Cellular structures appear at the site of ruptured ovarian follicles in a number of anamniote animals but, despite numerous investigations, we still have no conclusive evidence for an endocrine function of these structures. The mammalian corpus luteum may have been first concerned with making the mammary glands functional. Very few mammals have been investigated completely, but it seems that progesterone is essential for complete mammary lobulo-alveolar growth. A corpus luteum also seems to be capable of producing mammary gland development in the unmated female echidna, as described above.

Virtually nothing is known of the physiology of the monotreme corpus luteum. Successive changes of a secretory nature occur in the oviduct during passage of the egg, in synchrony with the development and regression of the corpus luteum. Frank Carrick and his coworkers found substantial quantities of progestagens and oestrogens in the peripheral plasma of a platypus that had intrauterine eggs with embryos at the 20-somite stage. The luteal cells looked active, and this suggests that the platypus corpus luteum may be functional throughout the period of intrauterine development. On the other hand, regressive changes are evident in the echidna corpus luteum by the time the embryo reaches the primitive streak stage some time before egg laying, and echidna corpora lutea 11 and 19 days after egg laying (approximately 1 and 9 days after hatching) are quite degenerate.

Both granulosa and theca interna cells contribute to the glandular tissue of the marsupial corpus luteum. Leon Hughes has shown that production of both the mucoid coat and the shell membrane of the ovulated egg are stimulated by the injection of oestrogen in the ovariectomized brush possum *Trichosurus*. The decline of the progesterone-induced secretory phase in the uterus of the North American opossum *Didelphis* begins on the 11th day of the 13-day pregnancy. Marilyn Renfree has repeated Carl Hartman's early experiments and found that embryonic development may continue to term in the uterus even if ovariectomy is carried out as early as day 6. Maximum progesterone secretion by the brush possum corpus luteum occurs on days 12 to 13 of the 17-day pregnancy and a decline is evident by day 15. The uterus does not become secretory if ovariectomy is carried out before day 8, but if the corpus luteum is removed later, embryonic development is not interrupted and females may give birth normally. Embryonic development also proceeds to term in those marsupials that have gestation periods approaching 30 days, if ovariectomy is performed after about day 7.

Hugh Tyndale-Biscoe has suggested that the critical factor in marsupials is the ability of the uterine endometrium to remain in a secretory condition without further ovarian stimulation, once it has been primed by secretions from the corpus luteum. Thus the marsupial corpus luteum may be an organ for the initiation of secretory changes in the uterus rather than for the maintenance of gestation. But whilst a few days' progesterone secretion by the corpus luteum may be sufficient to enable pregnancy to proceed to term in the brush possum, it is possible that kangaroos, with their longer gestation periods, may be dependent on substances produced by the developing embryo or fetus for pregnancy to proceeding to term after removal of the corpus luteum. This point is discussed later.

The blastocyst

The large mass of yolk in monotreme eggs is not partitioned between cells when cleavage begins; segmentation is confined to the embryonic area at one pole. Marsupial and eutherian eggs on the other hand undergo holoblastic cleavage, but the vestigial yolk body of marsupial early cleavage stages serves as one reminder of ancestors with yolky eggs, and early marsupial development is rather more reminiscent of that of monotremes than of eutherians. Marsupial and monotreme early blastocysts are not divisible into trophoblast and inner cell mass as is the eutherian blastocyst (Fig. 2-9).

The placenta

The best evidence of mammalian derivation from oviparous ancestors is the origin of the placenta from the extraembryonic membranes of oviparous amniotes (see Fig. 2-7).

We commonly think of the mammalian placenta as being chorio-allantoic since a portion of the chorion is vascularized by allantoic blood vessels. However the chorion can still act as a placental structure even if allantoic blood vessels never reach it. In all marsupials, part of the chorion is vascularized by vessels from the yolk-sac, and in most of the group vascular and non-vascular yolk-sac placentae constitute the sole means of exchange between embryo and uterus. However, two marsupial groups have independently evolved a chorio-allantoic placenta analogous to that of true mammals (Fig. 2-10).

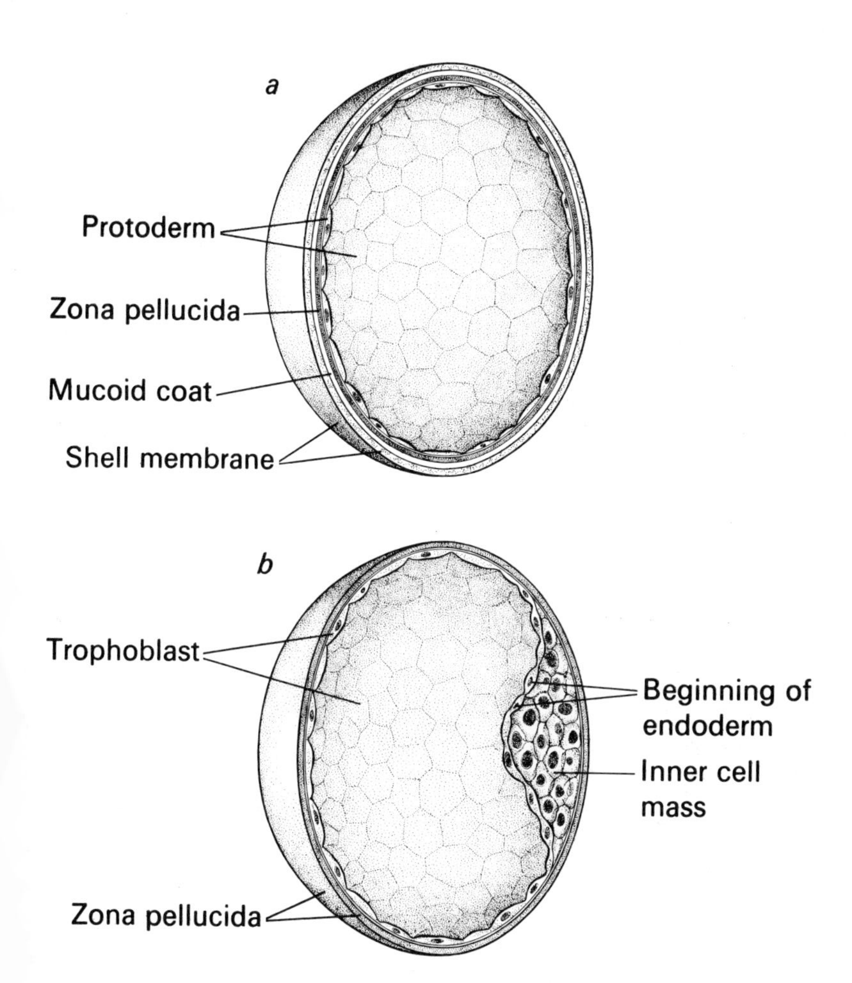

Fig. 2-9. Generalized drawings of halves of (*a*) marsupial and (*b*) eutherian mammal blastocysts. There is no inner cell mass in the marsupial blastocyst and the protoderm apparently has the potentiality to produce all types of both embryonic and extraembryonic tissues. The inner cell mass of the eutherian blastocyst gives rise to the embryo proper and at least the non-ectoderm portions of amnion, yolk-sac and allantois, whereas the tropoblast is essential for implantation and probably provides the cells of the ectodermal placenta.

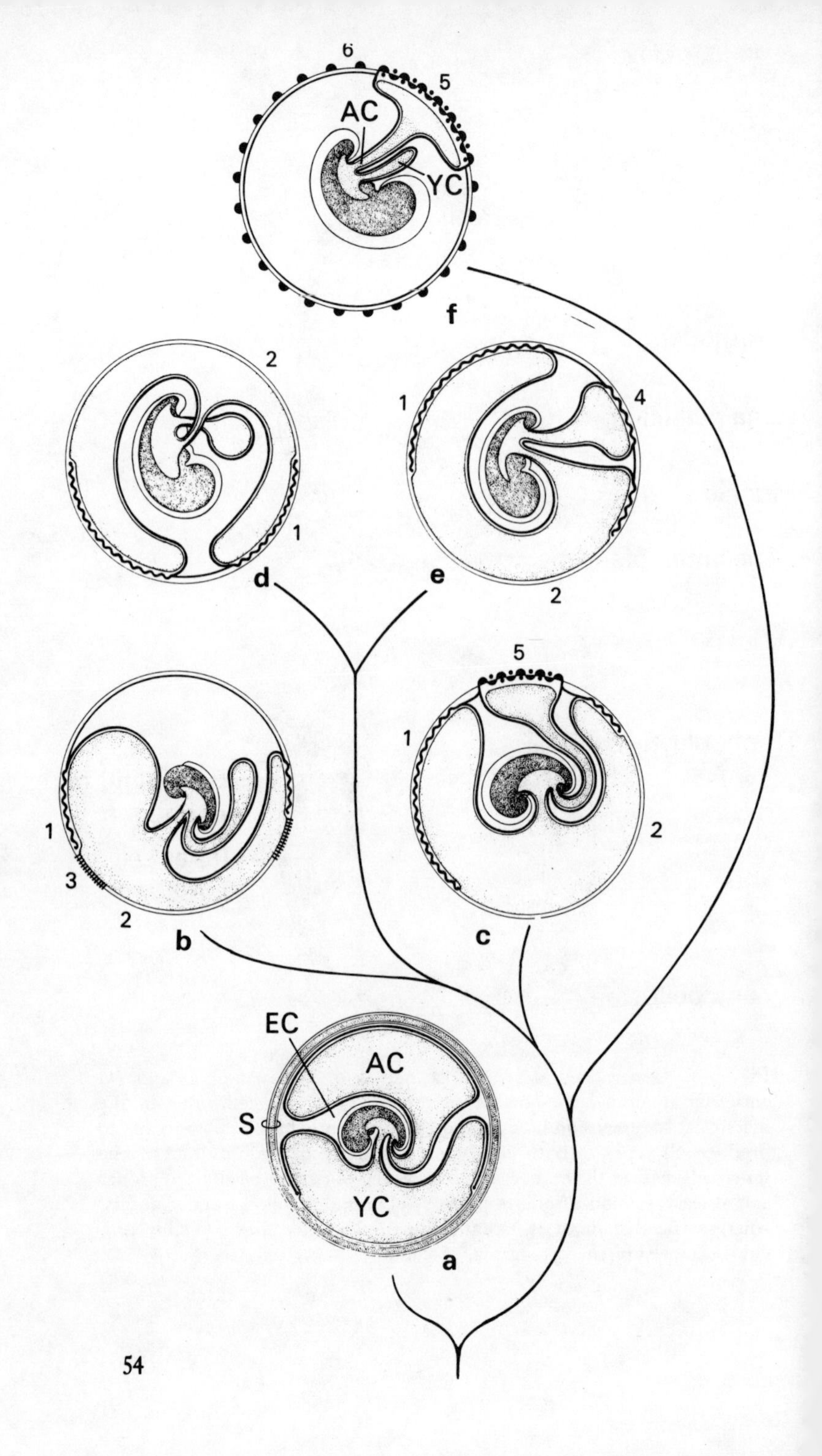

6
5
AC
YC
f
2
1
d
1
4
2
e
5
1
2
c
1
3
2
b
EC
AC
S
YC
a

Do marsupials 'recognize' pregnancy?

Since monotremes cannot distinguish between a fertilized and a non-fertilized egg in the uterus, we will chiefly be concerned here with the maternal recognition (or lack of recognition) of early pregnancy in marsupials.

In most eutherian mammals the luteal phase of the oestrous cycle is prolonged during pregnancy, and ovulation is usually suppressed. Maternal recognition of pregnancy takes a variety of forms (Book 3, Chapter 4), but, in general, the early embryo signals its presence at, or prior to, implantation and sets in train a complex series of interactions between uterus, ovary and pituitary. Marsupial gestation periods range in length from 12 to 37 days (Table 2-1), and although the longer marsupial gestation periods exceed the shorter eutherian ones, the largest neonatal marsupial still weighs less than a newborn mouse. Furthermore, no marsupial is known in which the presence of an embryo in the uterus interrupts the oestrous cycle, although the suckling stimulus from the newborn young has a profound effect.

Marsupials solved the problem of 'premature' birth by adaptations of the mammary gland, which both feeds and warms the poikilothermic young. Although we have already seen that lacta-

---

Fig. 2-10. Evolutionary end-products of the modification of amniote extraembryonic membranes as placental structures, the embryos being shown dark. *a*, Egg-laying mammal; *b–e*, various marsupials; *f*, eutherian (4-week human). 1, vascular yolk-sac placenta; 2, non-vascular yolk-sac placenta; 3, syncytialized yolk-sac placenta; 4, apposed allantoic placenta; 5, syncytialized allantoic placenta; 6, invasive trophoblast. AC, allantoic cavity; EC, extraembryonic coelom; S, shell; YC, yolk-sac cavity. In the native cat (*Dasyurus*, type *b*) the allantois reaches the chorion and then retreats from it without forming a placental structure; in bandicoots (*Perameles* and *Isoodon*, type *c*) a complex allantoic placenta is formed at the close of gestation and the yolk-sac placenta remains functional until the young are born; in kangaroos and phalangers (Superfamily Phalangeroidea, type *d*) the allantois may grow to a large size but remains enshrouded in folds of the yolk-sac wall and in the koala (*Phascolarctos*) and wombat (*Vombatus*) the allantois reaches the chorion forming an apposed allantoic placenta (type *e*).

TABLE 2-1. Gestation periods, oestrous cycle lengths and birth weight ratios in various marsupials compared to those of the sheep

Species	Gestation period (days)	Oestrous cycle length (days)	Ratio wt newborn: wt mother
Virginian opossum	13	29	1:8300
Long-nosed bandicoot	12	26	1:4250
Brush possum	17	26	1:7250
Dama wallaby	29	30	1:10000
Wallaroo	32	33	—
Red kangaroo	33	35	1:33400
Western grey kangaroo	30	35	—
Eastern grey kangaroo	36	43	—
Swamp wallaby	37	31*	—
Sheep	148	16	1:14

* Oestrus occurs during gestation and an egg on the non-gravid side of the reproductive tract may be fertilized. (Data from: J. H. Calaby and W. E. Poole, *International Zoo Yearbook* **11**, 5–12 (1971); and G. B. Sharman, in *Viewpoints in Biology*, **4**, ed. J. D. Carthy and C. L. Duddington, London, Butterworth (1965).)

tion is not dependent on the prior occurrence of a pregnancy, nevertheless suckling provides a very effective 'pregnancy recognition mechanism' in marsupials because it arrests cyclic ovarian function. Whether this is the only marsupial pregnancy recognition mechanism is still unknown.

Hormone secretion during pregnancy

Cedric Shorey and Leon Hughes have found no significant increase in ovarian vein or peripheral plasma progesterone concentrations in pregnant brush possums compared to their non-pregnant counterparts at equivalent stages of the oestrous cycle. There are not even any ultrastructural differences between corpora lutea or uterine secretory cells of pregnant and non-pregnant brush possums at similar postovulatory stages.

The only other marsupial in which hormone production during pregnancy has been investigated is the dama wallaby *Macropus eugenii*. Whereas the brush possum secretes postovulatory levels of progesterone comparable to those of sheep, cows and guinea pigs, the dama wallaby seems to be somewhat like the elephant (Book 4, Chapter 1) in that it produces very little progesterone. There is just a suggestion that female dama wallabies in mid pregnancy have more progesterone than non-pregnant females, but by the end of pregnancy the difference is swamped by rising preovulatory levels which are similar in pregnant and non-pregnant animals. Although pregnancy will proceed to term in ovariectomized animals, this also abolishes the mid pregnancy rise in progesterone, suggesting that the hormone is not placental in origin.

Embryotrophe

Uterine 'milk' or embryotrophe is produced by the uteri of both pregnant and non-pregnant marsupials, and its production appears to be controlled by cyclic ovarian secretions. The follicle or early corpus luteum exerts a unilateral effect, as Sandra von der Borch showed in the brush possum, so that in both pregnant and non-pregnant animals the uterus receiving the single ovulated egg is always heavier and produces more embryotrophe than the contralateral uterus. A similar unilateral effect is seen in the dama wallaby *Macropus eugenii*, although embryos transferred experimentally to the uterus on the opposite side to the corpus luteum develop normally. The embryo or membranes also cause a local stimulation of the endometrium, but this does not constitute maternal recognition of pregnancy in the eutherian sense, as the effect is independent of the corpus luteum and affects only the gravid uterus.

The pituitary gland in marsupial pregnancy

Experimental work on the pituitary control of marsupial reproduction has centred on the dama wallaby which Pat Berger has shown exhibits both lactational and environmentally controlled embryonic diapause (delayed implantation) (see Book 4, Chapter 1). John Hearn was able to show that the corpus luteum arrest that occurs during diapause is evidently due to inhibition by the pituitary and not to lack of gonadotrophins. Hypophysectomy not only results in a reactivation of the corpus luteum, but the structure then functions for its normal lifespan. Since all follicle growth is abolished in the hypophysectomized wallaby, the pituitary gland is presumably essential for follicular growth and ovulation, as in eutherian mammals.

In eutherian rats and mice which have lactation-controlled delayed implantation, hypophysectomy results in suspension of all ovarian activity and the blastocysts eventually degenerate.

Parturition in marsupials

Antigenic incompatibility between developing embryo and mother (discussed in Book 4, Chapter 4) can hardly be a problem in the monotremes, for the developing embryo is protected by the egg shell, itself of maternal origin, which remains intact throughout intrauterine gestation. Marsupial blastocysts and early embryos are likewise protected by the shell membrane which remains intact until at least the beginning of organogenesis.

There is good reason to believe that the fetus itself, and not its mother, exercises a dominant control over the time of parturition in eutherian mammals (see Book 2, Chapter 3). In marsupials this

TABLE 2-2. Behavioural evidence of 'pregnancy recognition' in the female red kangaroo during a 4-hour period at the close of gestation

Description of female	Time before expected parturition		No. of 5-minute periods during which female:		
	Days	Hours	Cleaned pouch	Adopted birth position	Licked urogenital area
Not pregnant (controls)	5	—	1	1	0
	3	—	3	2	1
Pregnant	2–3	—	0	0	0
Never mated, potential foster-mother*	1	—	3	0	0
Pregnant, gave birth at end of 4-hour period	—	4	44	22	16

* Newborn young of the female that gave birth was transferred to the pouch of this female and raised to maturity. (Data from G. B. Sharman and J. H. Calaby, Reproductive behaviour in the red kangaroo. *CSIRO Wildlife Research*, **9**, 58–85 (1964).)

concept could be tested by the transfer of fertilized eggs of one species to another with a longer or shorter gestation period. This has not been done, but Bill Poole's extensive observations on gestation periods in Eastern grey kangaroos carrying embryos sired by Western grey kangaroos, certainly points to a control of gestation length determined by the genetic constitution of the fetus. There is considerable development of the fetal adrenal cortex in late gestation in a number of marsupials, so perhaps it is responsible, as in some eutherians, for initiating the remarkably complex series of events that accompany parturition.

Although there is no eutherian-type pregnancy-recognition mechanism in marsupials during the early stages of intrauterine gestation, there can be no doubt that female marsupials 'recog-

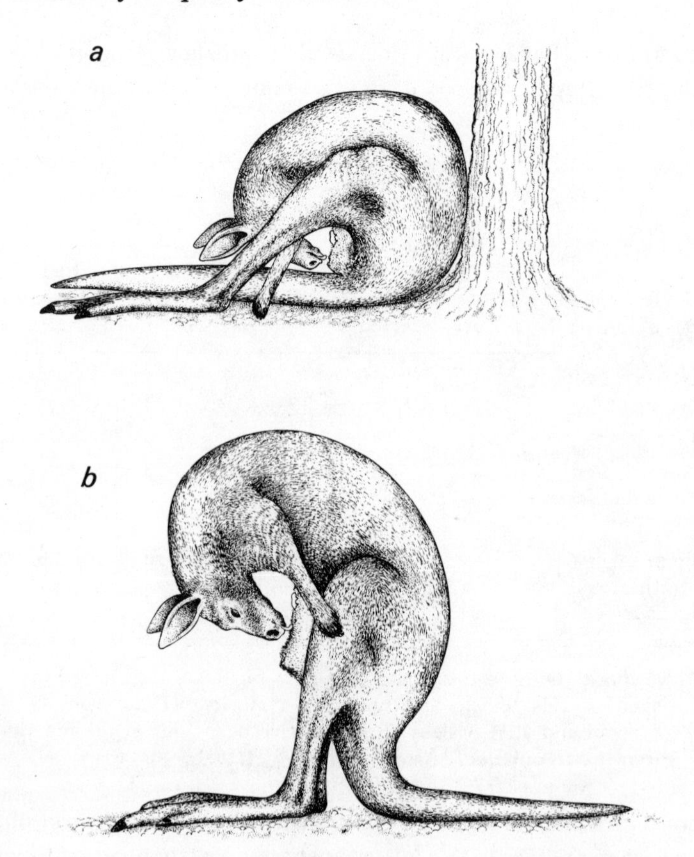

Fig. 2-11. Parturition in two species of kangaroos. *a* Red kangaroo, sitting birth position with back supported. *b* Eastern grey kangaroo, standing birth position. In both instances the mother is removing traces of membranes and birth fluids from the fur by licking behind the young as it crawls towards the pouch opening. (After G. B. Sharman, 'Gestation, naissaince et progression du jeune vers la poche marsupiale dans la famille des kangourous'. *Zoo: Société Royale de Zoologie d'Anvers* **34**, 124–32 (1969).)

nize' pregnancy during its terminal period. At this time, in the red kangaroo, the fetus induces an increased frequency of several patterns of maternal behaviour (Table 2-2) including pouch cleaning, grooming the fur and external urogenital area, and sitting in the 'birth position' with tail passed forward between the hind legs (Fig. 2-11).

Once the young has entered the pouch, and attached itself to a teat, the suckling stimulus inhibits further ovarian activity so that oestrus is prevented, or the development of the corpus luteum is arrested in those marsupials that carry a dormant blastocyst during pouch suckling.

POSSIBLE STEPS IN THE EVOLUTION OF MAMMALIAN VIVIPARITY

The monotremes have a somewhat more specialized reproductive system than other oviparous species, such as birds and reptiles. There has been some reduction in egg-yolk content, the developing embryo is retained in the reproductive tract for a longer time, and the corpus luteum regulates the uterine secretion of embryotrophe which nourishes the embryo inside its shell. There is also a period of extrauterine 'gestation' in the pouch-bearing echidnas.

The newly hatched monotreme young has both an egg tooth and a caruncle (Fig. 2-12). We believe that the egg tooth is used to slash open the membranous amnion, and the caruncle, in conjunction with movements of the forelimbs (Fig. 2-8), to break the shell at hatching. The monotremes are unique amongst amniotes in possessing both of these structures; oviparous reptiles only have one or the other.

The marsupials are more closely related to eutherians than to monotremes (Fig. 2-1), but they have retained some features of the prototherian reproductive pattern which were presumably characteristic of the common ancestors of both groups.

Assuming that in pre-mammalian species, the young were derived from internally fertilized, large yolked, cleidoic eggs, we can envisage the following steps in the evolution of viviparity in marsupials:

(1) The appearance of paired vaginae derived from the lower portion of the oviducts lying lateral to the ureters; the develop-

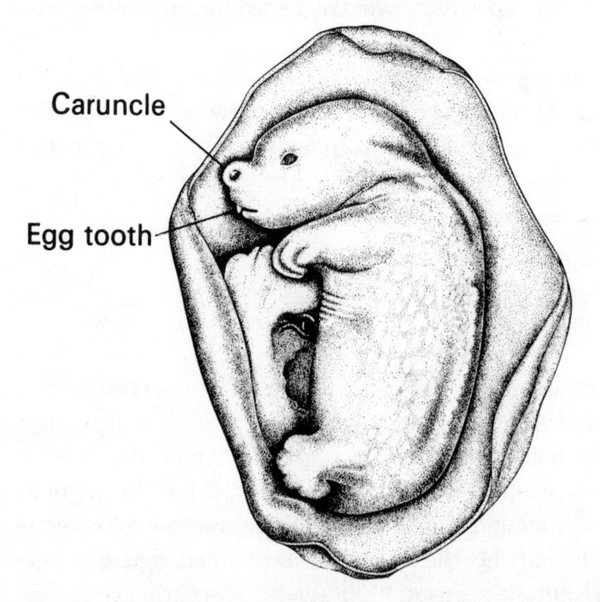

Fig. 2-12. An echidna just prior to hatching, dissected from the egg shell. (From J. P. Hill and G. R. de Beer, 'Development of Monotremata: VII'. *Transactions of the Zoological Society of London* **26**, 503–44, Fig. 41 (1950).)

ment of a median (pseudovaginal) birth canal; the virtual abolition of the cloaca in adult animals.

(2) Reduction of egg yolk coincident with evolution of more efficient intrauterine nutrition, and virtual abolition of the egg shell. Specialization of the yolk-sac and other extraembryonic membranes as organs for the exchange of substances between uterus and embryo.

(3) Endocrine control by the corpus luteum of the uterine secretion of embryotrophe, which nourishes the developing embryo (a mechanism shared with monotremes).

(4) The development of teats on the mammary glands (absent in monotremes) and adaptations of the mouth of the newborn, allowing it to become firmly attached to the teat during extrauterine 'gestation'.

These adaptations are common to all marsupials so far investigated, but there has also been some divergent evolution within the group. Most probably the ancestral marsupials did not possess a marsupium or pouch, the structure from which the group is named, and even today some small mouse- and rat-like opossums are pouchless. Pat Woolley has described a remarkable variety in the pouches of various members of the Australian carnivorous marsupials (Dasyuroidea), which suggests independent evolution of the pouch in these animals. In the bandicoots (Perameloidea) the pouch opening points backwards, in the kangaroos and possums it points forwards, and in wombats it is a central slit. The incubatorium or brood pouch of the echidnas is thought not to be homologous with the marsupium. In view of this variety, it seems likely that the pouch evolved on several separate occasions within the marsupials.

Once the marsupials had embarked along the evolutionary path characterized by 'premature' birth and pouch gestation, further evolution of intrauterine gestation was probably redundant. Birth of young of small size and their subsequent extra-uterine gestation in a marsupium is a very efficient device.

The only advantage of a longer gestation period would be that the young were better developed at birth, and hence better able to reach the pouch. Although this has happened in the kangaroos, they also have developed the longest period of pouch development.

Eutherians – the great leap forward

In mid-Cretaceous times around 100 million years ago, North America and Eurasia (collectively Laurasia) were united across what is now the Atlantic Ocean and at least a tenuous connection existed between Eurasia and North Africa. The old southern supercontinent of Gondwanaland had started to drift apart, although the present Antarctic, Australian and South American continents maintained a fairly close geographic relationship with one another, and Antarctica was further north than at present and

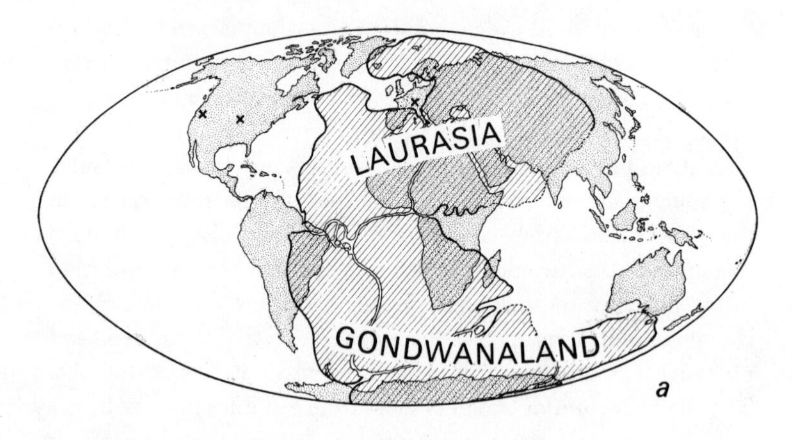

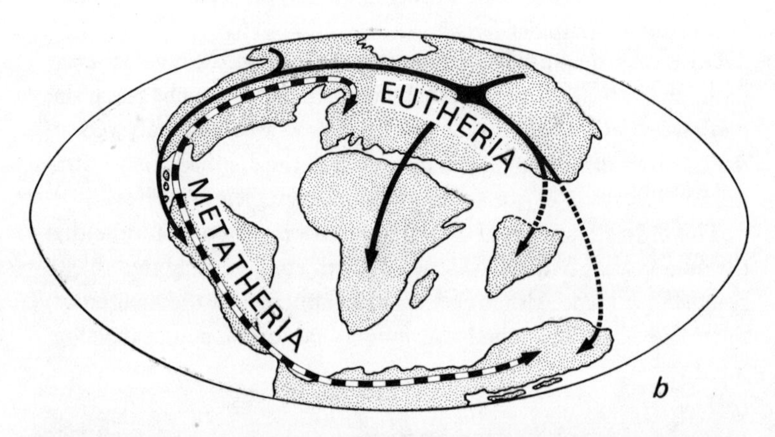

Fig. 2-13. The changing face of the earth and its effect on mammalian migration, distribution and evolution. In Triassic times (200 million years ago) the universal land mass, Pangaea, was made up of two supercontinents Laurasia and Gondwanaland (hatched in *a*). These are superimposed on the present-day distribution of the major land masses of the world (stippled in *a*). Fossil marsupials are found at the sites marked '×' in North America and Europe but no living marsupials occur there except for possible recently immigrated or transported Virginian opossums in North America.

The positions of the continents at the close of Cretaceous times (about 65 million years ago) are shown in *b*. Africa and India were no longer connected with South America, Antarctica or Australia, and South and enjoyed a milder climate. North and South America were not joined, although there may have been an archipelago between them. At about this time, and certainly before the close of the Cretaceous (65 million years ago) a therian stock of mammals split into two distinct groups which became the Metatheria (marsupials) and Eutheria (true or placental mammals) (Fig. 2-13).

More than twenty years ago Robert Hoffstetter said: 'Marsupials and Placentals (i.e. eutherian mammals) have too many analogies in their adaptive potentials for them to have developed simultaneously in the same region; it is likely that a geographical isolation mutually protected them from the beginning'. We can assume that when the marsupials and eutherians first came to occupy a common geographical area, as has been the case in several regions from late Cretaceous times onwards, they were already sufficiently specialized to be reproductively isolated from one another. Dick Tedford suggests that the common ancestors of marsupials and eutherians were a group loosely described as 'therians of metatherian–eutherian grade' which have been found as fossils in both North America and Asia and were possibly of world-wide distribution. Their mode of reproduction is unknown but they were terrestrial and hence amniote, and milk-fed their young. The most reasonable assumption is that this ancestral therian mammal stock was oviparous, with limited intrauterine development before egg-laying, and had young that were hatched at an early stage of development. In short, the basic therian pattern of reproduction was similar to that of the present-day

---

North America were probably tenuously connected by an archipelago. It is assumed that the eutherians may have evolved in Asia and thence spread throughout the world. The metatherians (marsupials) presumably evolved in isolation in North (or South) America and spread to Europe (where they became extinct in Miocene times) and, via Antarctica, to Australia and New Guinea.

(For further details of distribution of fossil and recent marsupials related to the changing face of the earth see R. H. Tedford, 'Marsupials and the new paleogeography'. In *Paleogeographic Provinces and Provinciality*. Ed. C. A. Ross. Society of Economic Paleontologists and Mineralogists Special Publication No. 21 (1974).)

monotremes. If the marsupial and eutherian mammals arose in separate geographical regions from a common oviparous ancestor, then these two modes of viviparous reproduction are an example of parallel evolution.

The few eutherians that have been studied in detail reveal a bewildering array of reproductive mechanisms which may be regarded as adaptations for viviparity. Eutherian mammals are grouped into four cohorts (Fig. 2-14). Very little is known about the physiology of reproduction in the Mutica, because whales are rather difficult to study, but we do know something about representative members of each of the remaining cohorts. In general, the evolution of viviparity in eutherian mammals has been characterized by an increasing dependence of the embryo and fetus on endocrine secretions from the maternal ovaries and the fetal placenta – and ultimately from the fetus's own endocrine glands.

All of the earliest fossils belong to the Order Insectivora, a group that persists to the present day, but unfortunately we still know little about their reproductive processes. The ferungulate line (Fig. 2-14) split from the basic mammalian stock at an early period in evolution and the carnivores had become a distinct group of primitive predators by the close of Cretaceous times. Most carnivores have remained monoestrous, and this may be a basic eutherian reproductive characteristic. Oestrus and ovulation generally occur only once in each breeding season, and are always followed by a period of ovarian, uterine and mammary activity that is similar in pregnant and non-pregnant females, so that a 'pregnancy recognition' mechanism (Book 3, Chapter 4) may be unnecessary. There appears to be no luteolytic influence of the uterus, and the developing fetus is dependent on hormone secretion by the maternal ovaries throughout most of gestation.

The ungulates (odd- and even-toed herbivores) are apparently the living representatives of an early offshoot of the carnivores. Like the carnivores, their fetuses are largely dependent on maternal endocrine secretions. They differ from the carnivores in being polyoestrous, and this has been made possible by the development of a uterine luteolytic mechanism (prostaglandin $F_{2\alpha}$) which

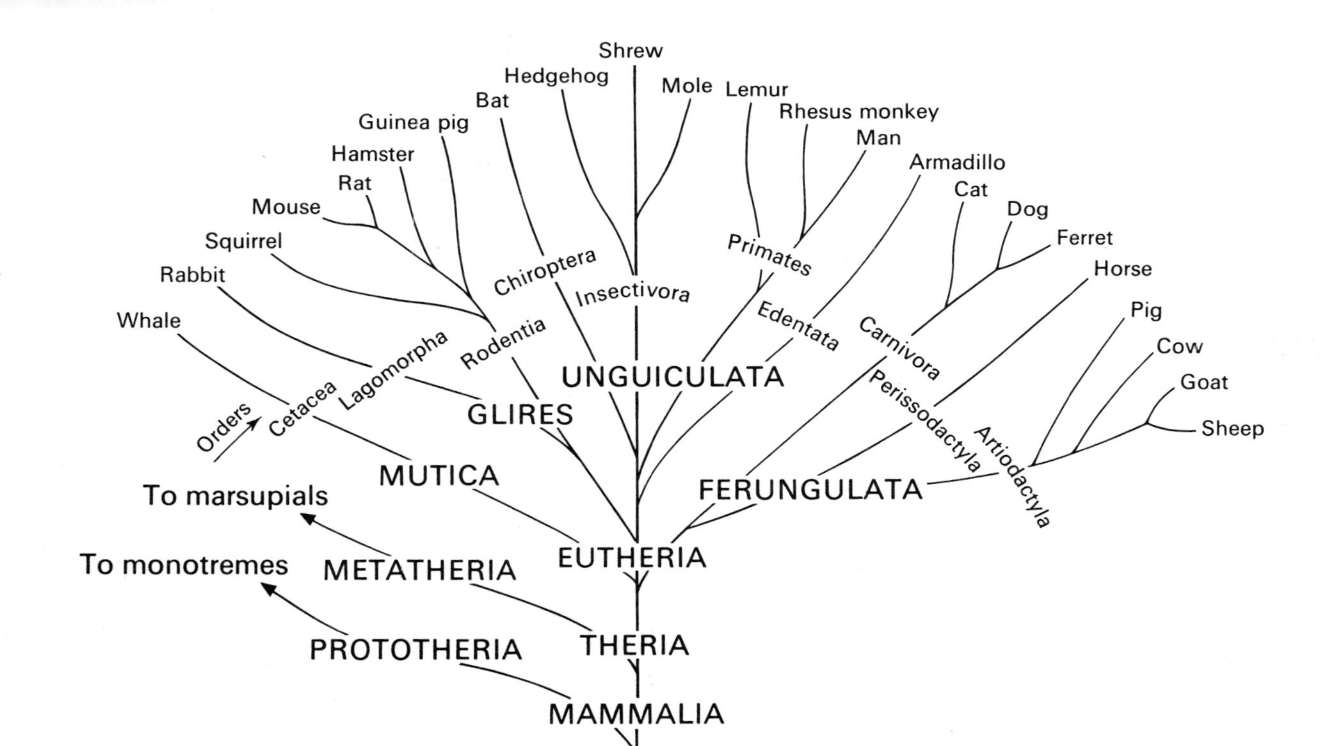

Fig. 2-14. Taxonomic, and probable evolutionary, relationships of the mammals in which reproductive processes are most commonly studied.

destroys the cyclic corpus luteum if pregnancy does not occur. The ability to return to oestrus rapidly in the event of a failure of fertilization or pregnancy has obvious selective advantages in animals that bear but a single offspring, as do many ungulates.

The primates are included in the Cohort Unguiculata (Fig. 2-14) and the evidence suggests that they are derived from an initial radiation of primitive primates in Palaeocene times. If the rhesus monkey and human female are to be regarded as representative examples, then primates solved the problem of becoming polyoestrous in quite a different way to the ungulates. The cyclic corpus luteum is apparently dependent on an initial burst of pituitary luteinizing hormone (LH) secretion which allows it to function for about 14 days before regression. In the event of a pregnancy, the placenta secretes a luteotrophic hormone, chorionic gonadotrophin, that can prolong the life of the cyclical corpus luteum.

In some of the Cohort Glires pregnancy recognition begins at mating; the stimulus of copulation is necessary to produce a fully functional corpus luteum, which has a similar lifespan in the pregnant or pseudopregnant animal. The uterine luteolytic mechanism, although it exists, is relatively unimportant. As might be expected the developing embryos and fetuses are almost entirely dependent on maternal hormones throughout pregnancy.

A wide array of reproductive mechanisms has been described in the comparatively few eutherian mammals thus far studied. The only conclusion that we can draw is that protection of the developing young has been of such paramount importance that many alternative solutions have been perfected. The most intriguing differences are revealed by study of closely related species. For example, the sheep placenta produces progesterone and the ewe can safely be ovariectomized after mid gestation without disturbing the pregnancy, whereas the goat has no placental progesterone and depends on the maternal ovaries throughout gestation. What was the situation in the common ancestor of these two forms of domestic animal? We do not know, but the answer may be found if more representatives of the group to which both sheep and goats belong are studied. Since there are only a limited number of variables to experiment with in eutherian reproductive mechanisms, we must not be surprised to find numerous instances of parallel evolution in different orders, and divergent evolution in closely related species.

The main advantages of viviparity undoubtedly lie outside rather than inside the uterus. The key to the success of mammalian viviparity has probably been lactation which provides the essential bond to unite mother and offspring thus making possible an extended period of postnatal growth and development. But viviparity was not achieved easily even after suckling made the rearing of underdeveloped neonatal young possible, for it also required extensive redeployment of internal organs. There were pitfalls along the way, as illustrated by the marsupial system where the ureters enter the base of the bladder, as in true mammals, but retain the 'reptilian' position between the two oviducts. The shifting of the ureters to the outside of the embryonic oviducts allowed the eutherian reproductive system to be modified and this, and other reproductive innovations, enabled man and his eutherian relatives to become the virtuosos of viviparity.

SUGGESTED FURTHER READING

Morphological studies on implantation in marsupials. R. L. Hughes. *Journal of Reproduction and Fertility* **39**, 173 (1974).

Observations of the comparative anatomy and ultrastructure of mammary glands and on the fatty acids of the triglycerides in platypus and echidna milk fats. M. Griffiths, M. A. Elliott, R. M. C. Leckie and G. I. Schoefl. *Journal of Zoology* **169**, 255 (1975).

Observations on the attachment of marsupial pouch young to the teats and on the rearing of pouch young by foster mothers of the same or different species. J. C. Merchant and G. B. Sharman. *Australian Journal of Zoology* **14**, 593 (1966).

Cyclical changes in the uterine endometrium and peripheral plasma concentrations of progesterone in the marsupial, *Trichosurus vulpecula*. C. D. Shorey and R. L. Hughes. *Australian Journal of Zoology* **21**, 1 (1973).

Intrauterine development after diapause in the marsupial *Macropus eugenii*. M. B. Renfree and C. H. Tyndale-Biscoe. *Developmental Biology* **32**, 28 (1973).

Reproductive physiology of marsupials. G. B. Sharman. *Science* **167**, 1221 (1970).

Gestation period and birth in the marsupial *Isoodon macrourus*. A. G. Lyne. *Australian Journal of Zoology* **22**, 303 (1974).

Control of reproduction in macropod marsupials. C. H. Tyndale-Biscoe, J. P. Hearn and M. B. Renfree. *Journal of Endocrinology* **63**, 589 (1974).

Comparative Biology of Reproduction in Mammals. Ed I. W. Rowlands. Symposium No. 15 of the Zoological Society of London. New York and London; Academic Press (1966).

Early Mammals. Ed. D. M. Kermack and K. A. Kermack. New York and London; Academic Press (1971). (Especially articles by Hopson and Clemens.)

Biogeographical considerations of the marsupial-placental dichotomy. J. A. Lillegraven. *Annual Review of Ecology and Systematics* **5**, 263 (1974).

3 Selection for reproductive success
P. A. Jewell

Darwin's theory of natural selection rests on the concept of the struggle for existence and the survival of the fittest. In every generation of animals a high proportion die or are killed before they are old enough to breed. Reproductive success is one component of general fitness. But of those individuals that do breed some contrive to leave more offspring than others. Not that the sheer production of many offspring is, in itself, an adequate measure of selective advantage: it is the offspring that themselves survive to breed that count.

When he wrote of a fit individual, Darwin meant one that was well adapted in every sense to its total environment, and this adaptation was maintained by natural selection. There are some attributes of animals, however, that do not seem to be necessary for survival in this way. The best examples are the bizarre adornments of the males of many species, such as the extreme plumage of peacocks or the extravagant antlers of deer. These Darwin saw to be of use only in gaining mates and he invoked the concept of sexual selection to explain their evolution. He argued that two major selective forces had given rise to these characters. Firstly, there was fighting amongst males for the possession of females; this, he supposed, had induced the evolution of secondary sexual characters that were useful in battle. Secondly, females could be highly discriminating between males, in accepting one as a mate, and this brought female choice into play.

The display organs of the males of some species, developed to excess, are sometimes described as 'non-adaptive'. I do not believe that this can be so, as all such characters are subject to the two quite distinct modes of natural selection that Julian Huxley called *survival selection* and *reproductive selection*. A species can tolerate some excess of sexual dimorphism so long as its exploita-

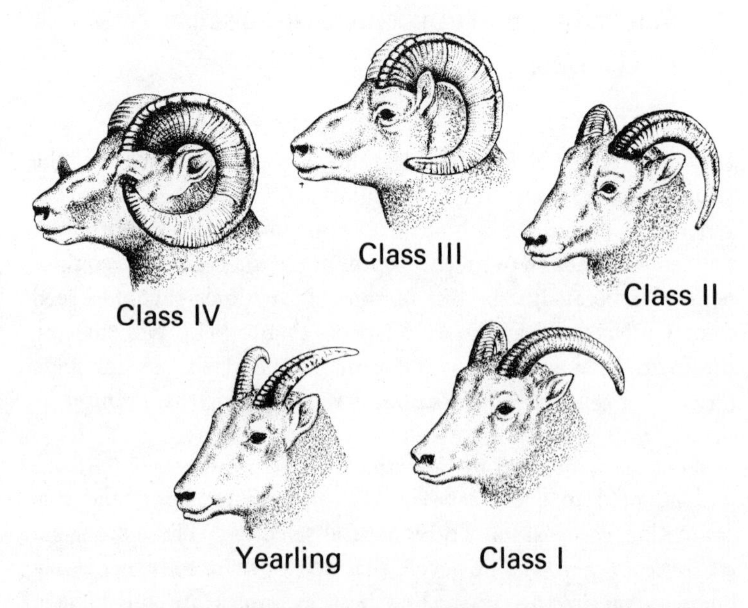

Fig. 3-1. Adult bighorn rams *Ovis canadensis* can be put into four horn-size classes. Class I are 2–3 years old; II, 3½–6; III, 6–8; IV, 8–16. Age can be determined in the field by counting the growth segments. (After V. Geist. *Canadian Journal of Zoology* **46**, 900, Fig. 1 (1968).)

tion of a particular ecological niche and its population density are maintained at optimal levels. This necessity to balance selective forces places the species (or the populations that comprise its effective adaptive units) on a knife-edge of compromise. Two examples will illustrate the dilemma. Firstly among birds, the male of the great-tailed grackle *Quiscalus mexicanus* enjoys a reproductive advantage from the attractiveness in display of his large tail feathers, but this same anatomical feature makes it difficult for him to fly in strong winds. As a result the male bird has to put up with a reduction in the time suitable for flying, his movements in the feeding area are restricted, and food intake falls. The dire consequence of this, and other stresses, is that the mortality rate amongst males is about twice that amongst females (such a disproportionate mortality is not at all unusual in animal

Fig. 3-2. Butting between two large (class IV) rams. (After V. Geist. *Mountain Sheep*, p. 203, Fig. 9. University of Chicago Press (1971).)

populations). The second example is in a mammal, the bighorn sheep *Ovis canadensis*, which was studied by Valerius Geist. In this species, very large horns in the male (Figs. 3-1–3-3) confer an advantage both in dominance and in reproductive success, but this in turn depends on a rapid growth rate which makes heavy metabolic and nutritional demands on the animal. Males that capitalize on rapid growth do not live as long as those that mature more slowly (see Fig. 3-4). The knife-edge balance between natural and sexual selection is elegantly displayed here and one can appreciate the intricate strategy of lifemanship.

The best way for an animal to become a successful breeder will depend upon the ecology of the species. The main ecological influences upon reproduction are: the food resource and competition for it, the availability of suitable sites in which to rear young, social organization, and the mating system and associated group reproductive behaviour. The components, and consequences, of selection for reproductive success that I want to consider are the differences between males and females, monogamy and polygamy, dominance and territoriality, and some aspects of

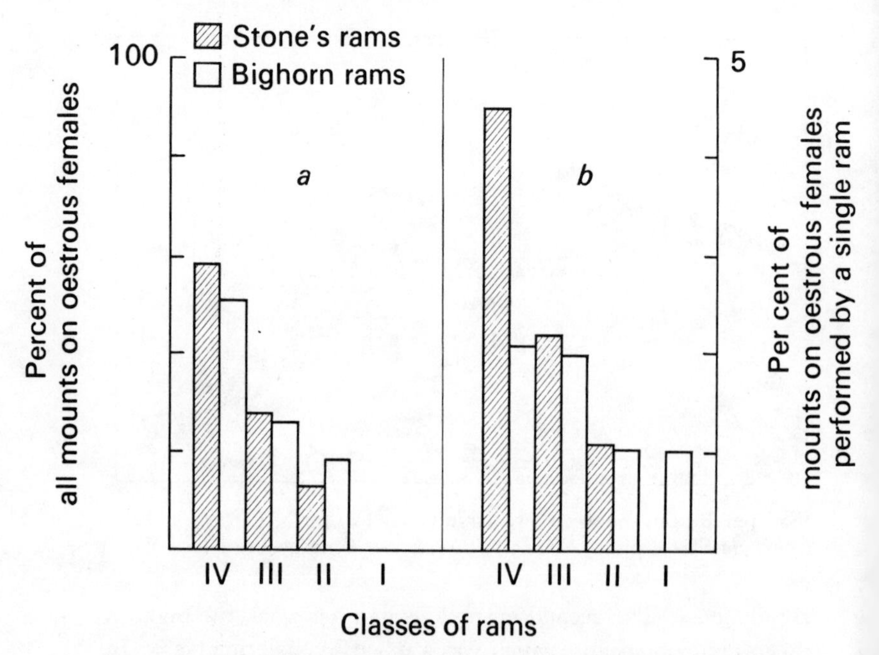

Fig. 3-3. The mating activity of American mountain sheep. Stone's sheep *Ovis dalli* are a race that inhabit the Cassiar Mountains in British Columbia and provide an example of a vigourous population. The bighorns *Ovis canadensis* were of poorer quality and were living at higher densities in southern Alberta. *a* Rams with larger horns more often mounted the oestrous females (for classes see Fig. 3-1). *b* Individual Stone's rams in class IV mounted about twice as many ewes as those in class III. Among the bighorns, class IV and III rams mounted equally often; the larger number of class IV present accounted for their being more successful as a group, as shown in *a*. (After V. Geist. *Mountain Sheep*, p. 215, Fig. 28. University of Chicago Press (1971).)

sexual selection. This discussion will be confined to mammals, but occasionally examples must be drawn from birds because they illustrate some phenomena about which we have little information in mammals.

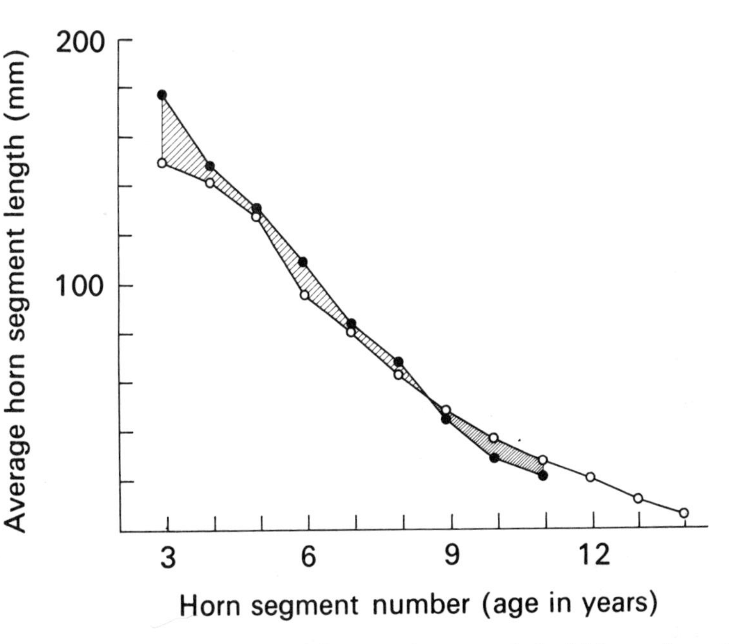

Fig. 3-4. Horn growth differences in rams of the bighorn sheep *Ovis canadensis* that died before and after the average age at death, namely 10–12 years. Horn segment growth between 3 and 8 years of age was greater for rams that died early, before average age of death (solid circles), than for rams that lived longer (open circles). (After V. Geist. *Mountain Sheep*, p. 182, Fig. 26. University of Chicago Press (1971).)

DIFFERENCES BETWEEN MALES AND FEMALES

A mating system is a framework of social relationships within which the individuals in a given population of a species find and compete for a mate. The mating system of a species is also an expression of ecological adaptation that greatly influences the fitness of individuals. The best strategy for males is often quite different from that for females, and the difference is most strongly shown in polygamous as opposed to monogamous mating systems. (Polygamy is a general term for having more than one partner of the opposite sex; polygyny implies more specifically a male consorting with and mating more than one female, and polyandry a female consorting with and mating more than one male (See Fig. 3-5.) The reproductive process in mammals, involving viviparity and sometimes a long pregnancy, often producing only a single offspring, means that the number of young that a female can produce in her lifetime is very limited. Also, in contrast to other vertebrates the female's role is protracted, since the newborn young are fed on milk necessitating a mother-based family unit and a high maternal energy expenditure. In comparison, the male of most species is fancy-free. A single male can sire a large number of offspring for whom he has few if any nutritional responsibilities, and the nutritional demands upon him for producing spermatozoa are negligible.

The female

The female can maximize her chances of passing on genetic material in a number of ways. Longevity is one: by adapting well to the demands of survival selection she can live through several breeding seasons and so produce many offspring in her lifetime. Even so a female large herbivore, for example, may produce a maximum of only ten or so youngsters. Alternatively, litter size can be increased and the females of many short-lived small mammals increase their productivity in this way. (The optimum litter size, it should be noted, is not necessarily the largest but is the one that contributes the greatest number of recruits to the next generation.) Early puberty may increase output, but adverse effects on the female's growth rate and longer-term fitness will limit this advantage.

Competition between females is a subject that has received little study, compared with that between males, but it can take a variety of forms that ultimately have an effect on reproductive success. Aggressive behaviour between females in a group gives rise in many species to female hierarchies which in turn may control access to limited food resources. This can be seen in species such as reindeer, eland and buffalo. In some mammals the females may be territorial, an attribute more frequently associated

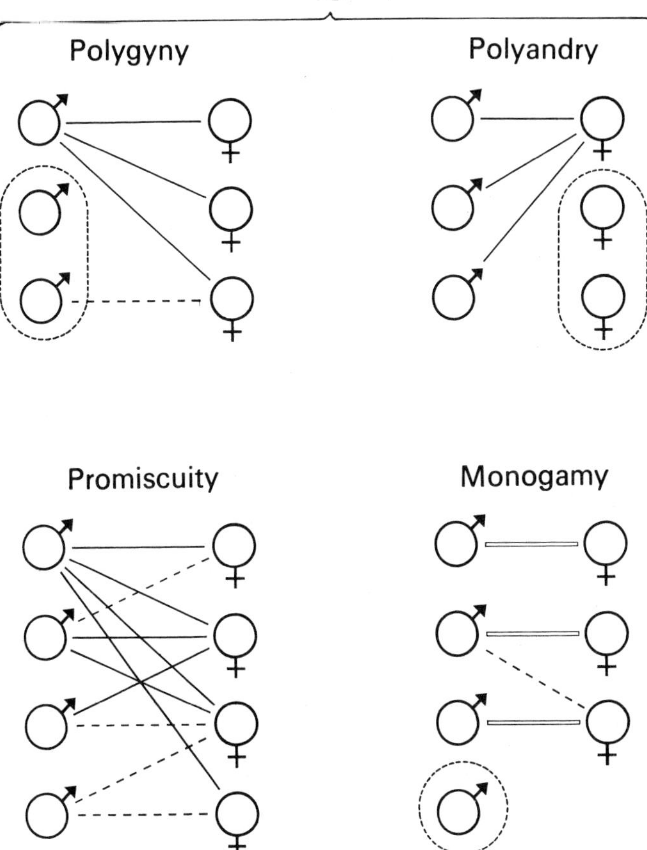

Fig. 3-5. Types of mating systems in mammals (and birds). Lines indicate successful mating attachments; dashed lines indicate attempts at copulation, not necessarily successful. Polyandry is included for completeness but does not occur in mammals except in certain human societies: it is quite common in birds where a male can incubate a clutch of eggs and rear the young. Note that subordinate males may attempt to intrude into polygynous systems and that copulatory activity is not equally shared in promiscuous systems. For monogamy the open bar is intended to suggest pair-bonding lasting more than one breeding season. Excluded males (or females) are indicated by a dotted oval or circle (Modified from Jerram Brown. *The Evolution of Behaviour*, Fig. 8-2. New York; Norton (1975).)

with the male. Research workers in Poland have found that, in certain circumstances, female bank voles *Clethrionomys glareolus* are territorial, and that adult females that are breeding prevent younger females from successfully gaining a territory. As a result the younger females fail to become sexually mature. The holding of territory is thus a prerequisite of breeding success. Many female insectivores, like the shrews and the mole, are territorial and so, probably, are most so-called solitary mammals, such as the lorisoid prosimians like the angwantibo *Arctocebus calabarensis*, and the notoriously aggressive female hamster *Cricetus cricetus*. 'Solitary' is not an accurate description for the way of life of such mammals because contact between individuals is probably frequent. They may certainly be described as asocial, but it is better to think of them as having a highly dispersed society.

Reproductive success does not necessarily involve direct competition, and this is particularly true of females. On the contrary close conformity to a pattern may be involved and this can be seen when females co-ordinate the time of parturition to establish a very short season of births. An example of this in African antelopes will be given later.

The male

The male mammal in most species is not involved in rearing the young, but has very great opportunities for passing on heritable characteristics. One male may inseminate a large number of females in a single season. Longevity alone is no longer a condition for high reproductive output (although if age is linked with some characteristic of dominance it will be). Obtaining access to breeding females is the prime criterion, and the major obstacle to achieving this objective is competition from other males. Competition between males has proved to be a potent selective force in evolution that has led to their large size compared with females, and to the enhanced development of those anatomical features and behavioural characteristics that either confer dominant status on an individual or enable it to obtain and hold a territory. A less common form of male competition is the slaughter by a newly instated dominant male of the existing youngsters of his predecessor. Examples of these systems will be given below, but first more needs to be said in general about monogamy and polygamy.

MONOGAMY AND POLYGAMY

The special role of the female mammal in bearing and rearing the young has led to a situation in the Mammalia in which monogamy is rare and polygamy or promiscuous mating by the male is common. This is in contrast to birds in which monogamy, or at least seasonal pair formation, is the usual form of mating system. The reason for this is clear: in mammals the male is usually not required to make any *direct* contribution to feeding and caring for the young, although he may, of course, play an important role in their survival. In many birds, on the other hand, both adults tend and feed the young, and the efforts of both parents are essential for reproductive success.

Monogamous mammals

Good examples of monogamy are seen amongst some of the terrestrial carnivores. Wolves, coyotes, jackals and foxes may all form pair bonds that can last for many years, and a similar situation is seen in some mustelids and viverrids. Capturing prey and bringing it back to the den is clearly analogous to the foraging activity of nesting birds. Similarly, the need to house-build may call for close parental co-operation, and the beaver is unusual amongst rodents in that the adults live in pairs, accompanied by juveniles. The families are self-contained, although several will combine together to construct and occupy a lodge.

The mammals that we know most about from field studies are the primates. Because of their close relationship to man, their social organization has been intensively studied over the last two decades. Even so, very few species have been found to be monogamous and to live in nuclear family groups. The South

American marmosets form pairs whilst rearing the young, and the male often carries the infants on his back, passing them to the female only for nursing. Another South American monkey that forms pairs is the titi *Callicebus moloch*, but the fidelity of the males is much in doubt!

The most clearly identified monogamous primates come from the evolutionary extremes of the order; amongst the 'advanced' apes there is the largest of the gibbons, the siamang *Symphalangus syndactylus*, which has recently been studied by David Chivers, and amongst the 'primitive' prosimians is the indri *Indri indri*, a lemur which Jon Pollock has recently studied in Madagascar. There are records of gibbons living together in the wild as a pair for at least seven years and indri for three years. Probably many are faithful to their partners for a lifetime.

All the monogamous primates have several features in common. The pairs vigorously defend a small territory in the forest, often by displaying at the borders and by noisy calling. There is no sexual dimorphism; the male and female are the same size, and either may be the dominant member of the pair. The birth interval may be long, only one infant being born every 2–3 years, infant carrying is well developed in both parents, and juveniles stay with their parents until they are quite old.

All the monogamous primates live in dense evergreen rain forests, and this common ecological factor must presumably favour the development of the nuclear family as a social unit. In such forests food is in steady but limited supply; the trees fruit and grow new leaves in a slow cycle. The food resources therefore tend to be well scattered throughout the territory, and under such conditions a single pair can evidently co-operate efficiently to rear their young. The male may find the food, but he permits the female and her youngster to have first pick. The male also offers protection when the group is threatened, by putting himself between them and danger. This is particularly important at the time of weaning when the youngster is learning to move independently and is vulnerable to accidents. The nuclear family, in its small but totally familiar territory, also provides an enduring environment in which juveniles may be able to survive successfully even after the death of a parent.

Despite all these painstaking field studies, however, we must admit that the advantages of monogamy in primates (and other mammals) still remain highly speculative. Unfortunately, little is known of the selective forces that act differentially on the two sexes in these species. It is also not known whether the rearing success of pairs that remain together for a long time is superior or inferior to that of more temporary partnerships.

The advantages of monogamy in birds

Because of the paucity of hard facts about the advantages of monogamy in mammals it is worth looking to birds for observational data. Studies of the kittiwake *Rissa tridactyla* by John Coulson provide information of exceptional interest. In this monogamous species there is virtually no sexual dimorphism and mortality rates in the two sexes are about the same. The birds are long-lived and tend to form life-long pair bonds. The pair separates temporarily at the end of each breeding season but reunites at the beginning of the next one. Both sexes assist in the incubation of the eggs and the care of the young. This joint experience ensures better co-ordination of parental responsibility, and can obviate loss of time at many stages in the breeding cycle. Some kittiwakes do separate and take new partners, however, even though both live and remain in the same colony. A divorce has occurred. Does this breakage of the pair bond lead to reduced reproductive success? It does. Birds that remain in the same mated pairs do show advantages over birds that divorce; on average they lay more eggs each year and they successfully rear a higher percentage of the young hatched. As a result of these two factors they fledge a significantly higher number of young (11.2 per cent more).

The data gathered by Coulson, that substantiate these assertions, are presented in Tables 3-1 and 3-2. A further word of explanation about breeding success in kittiwake colonies must be

TABLE 3-1. The mean number of young fledged per pair of kittiwakes according to breeding experience of the female, pair status and position in the colony

	Breeding experience (years)			
	2–4		5–16	
	Edge	Centre	Edge	Centre
Same mate	1.37	1.43	1.51	1.62
Divorced	1.23	1.31	1.34	1.47
Mate died	1.15	1.29	1.22	1.27

Sample sizes for 'same mate' and 'divorced' females were between 44 and 150; for 'mate died' they were between 16 and 26. (From J. C. Coulson. *Proceedings of the XV International Ornithological Congress*, pp. 424–33, table 9. Leiden; E. J. Brill (1972).)

added and this concerns the position of a pair in the colony. Those in the centre exhibit several advantages over those at the edge. The central birds have rather greater body weights and higher survival rates; a higher proportion of them breed successfully, and they are also superior in clutch size and number of young fledged. The factor of nesting position in the colony is obviously of particular interest in itself in relation to breeding success, but it will not be elaborated further here. It does, however, compel us to record data separately for birds at the centre and the edge, and Tables 3-1 and 3-2 have been set out accordingly; information for older, more experienced birds is also separated from that for younger birds. Evidently there is a marked advantage in the pair staying together and this is lost on divorce. It seems that a major factor provoking divorce is failure to rear young; sometimes an unsuccessful pair will try to breed again in a succeeding season, but after a second failure they inevitably part.

But divorcees are by no means the least successful breeders. Coulson's fascinating studies revealed that birds that had changed

TABLE 3-2. Differences in young fledged per pair of kittiwakes according to conditions. The differences are statistically significant

Age. Old parents produce	8.3%	more fledgelings than young parents
Position. Centre nests produce	7.3%	more fledgelings than edge nests
Mate. Fidel pairs produce	11.2%	more fledgelings than divorcees
Mate. Fidel pairs produce	19.2%	more fledgelings than pairs in which a mate has died

(From J. C. Coulson. *Proceedings of the XV International Ornithological Congress*, pp. 24–33, table 10. Leiden; E. J. Brill (1972).)

their partner because of the death of their previous mate, had a significantly poorer reproductive performance than divorced birds. There was no apparent reason for this difference, and its significance is enhanced by the fact that divorce was all the more likely to have occurred amongst potentially poor breeding birds. It is tempting to make a comparison between our society and that of the kittiwake, and suggest that the effects of voluntary divorce are disruptive but acceptable, whereas the death of a partner is a traumatic experience.

Polygamy in mammals

Turning to the system more usually found in mammals, that of polygynous mating, it is the male that exhibits the striking evolutionary consequences of intrasexual competition. Nowhere is this more clearly seen that in the Order Pinnipedia that exhibits extreme sexual dimorphism and gregarious breeding (Fig. 3-6). The remarkable similarities in social structure between species of sea-living mammals suggest that a few overriding ecological factors, peculiar to an amphibious habitat, have moulded their main reproductive strategies. A model for the evolution of pinniped polygyny has been developed by George Bartholomew, and is presented here in Figure 3-7. This may look complex but it is worth careful study.

Fig. 3-6. Challenge for harem mastery. Bull elephant seals on their territories at the rookery.

Two key attributes, shared by no other mammals, dominate pinniped biology: they exploit offshore marine food resources, and they copulate and give birth on land (large circles, Fig. 3-7). They have, in addition, developed certain general mammalian features to a high degree (small circles). These take the following forms:

(*a*) The animals display large size and much subcutaneous fat which contribute to heat conservation, and the fat provides an energy source in the absence of food. Mammals that are large and have extensive fat deposits can go for long periods without food because their ratio of surface area to rate of metabolism is much more favourable than in small mammals (low weight-relative metabolism; see Fig. 3-7).

(*b*) One male has the ability to fertilize many females.

(*c*) There is high male aggressiveness and sex drive, in part controlled by testosterone.

(*d*) Aggressive behaviour amongst males results in the estab-

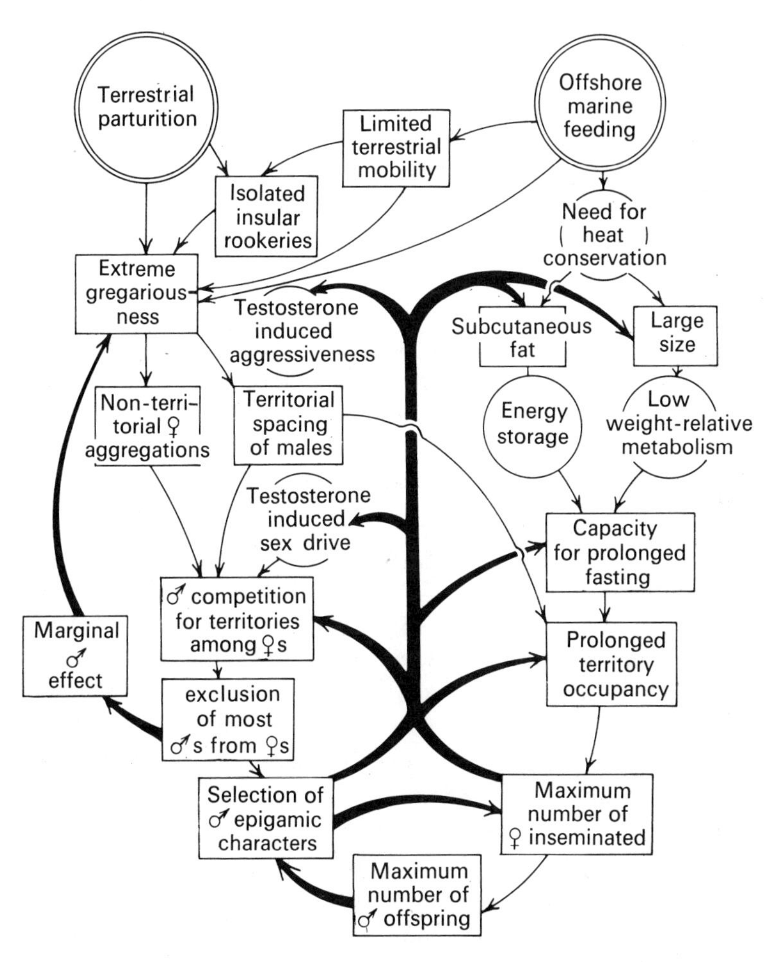

Fig. 3-7. Diagram showing the interaction of factors underlying the evolution of polygyny in pinnipeds. The two large circles indicate two distinctive features of pinnipeds, while the small circles name characters common to most mammals. The properties described in rectangles are those characteristic of polygynous pinnipeds. Direct interactions are shown by thin arrows, and positive feedback effects by broad arrows. (From G. A. Bartholomew. *Evolution* **24**, 546, Fig. 1 (1970).)

lishment of a dominance hierarchy, or territories, or both (Fig. 3-6).

(*e*) Males are free from nutritional responsibility for their offspring.

It is Bartholomew's thesis that, given marine feeding and terrestrial parturition, the general physiological and behavioural characteristics of eutherian mammals will lead to polygyny and this permits the congruent organization of many otherwise disparate attributes (rectangles in Fig. 3-7). For example, both male and female pinnipeds are pre-adapted to abstinence from food. The females are able to lactate and can at the same time stay close to the male until they come into oestrus, whilst the male can engage in vigorous combat and herding. Gregarious breeding is advantageous because it brings together animals that are normally dispersed over enormous areas in search of food. The very cumbersomeness of the animals on land leads to crowding and intense aggression between males, so that most males are effectively excluded from breeding. The aquatic way of life, leading to very large size, has endowed the males with a grossness that they exploit as an epigamic character (one that attracts or stimulates individuals of the opposite sex during courtship; see Fig. 3-7).

In terms of reproductive success the repercussions of the pinniped way of life are dramatic, at least for males. In the case of the Northern fur seal, for example, it can be calculated that a territory-holding male could easily produce eighty male offspring in his lifetime (assuming an average harem size of forty, an 80 per cent conception rate, breeding for five seasons, and a 1 : 1 sex ratio amongst pups). A male that failed ever to establish a territory might produce none.

Many large terrestrial herbivores live in herds and the males do not hold territories, but they may live apart from the females for much of the year. When they mingle with the female herds, the males have distinct roles and form their own rank hierarchy. Naturally, the dominant males could be expected to do most of the mating; this has been observed to be the case, but there have been few studies that have actually quantified this success. Clear evidence is available from Geist's study of the Canadian bighorn sheep *Ovis canadensis* (Fig. 3-3), but more detailed information was obtained in a study that my colleagues and I conducted on the feral Soay sheep on the isolated island of St Kilda, far off the west coast of Scotland.

Soay sheep resemble the wild mouflon, and show many primitive anatomical and behavioural characters. They are small, and whilst the males have robust horns they do not compare in magnificence with those of the Canadian bighorn sheep (Figs. 3-1 and 3-10). Consequently, combat between Soay males is conducted more by kicking displays, body wrestling, head rubbing and submissive behaviour, than by horn-clashing.

By recording the victor of such encounters, a rank order could be assigned to each Soay ram. There was found to be a linear hierarchy with each ram in the sequence dominant to those below it and subordinate to those above. This hierarchy is set out in Fig. 3-8. The period during the autumn when most of these interactions occurred is shown in Fig. 3-9. The sorting-out of rank was largely determined by agonistic behaviour (Fig. 3-10) that occurred before the main peak of oestrus amongst the ewes, although some ewes were showing oestrus at this time without conceiving. By the time of the first fertile oestrus the rams knew their own rank status and subordinates gave way to their superiors. As a result, a dominant ram could indulge in a short period of consortship with a receptive ewe, and during this 'tending bond' (a contracted remnant of pair bonding?) he inseminated her and presumably sired her lamb. Fig. 3-8 presents a measure of the reproductive success attributable to dominance. In order to maximize his success a ram should not waste unproductive time with a ewe once she has conceived. Conception will result from insemination during the first day of the ewe's oestrous period, so time spent with a ewe when her oestrus extends into a second day is 'wasted' as far as the reproductive success of the ram is concerned. Fig. 3-8 shows a clear decrease in the duration of the

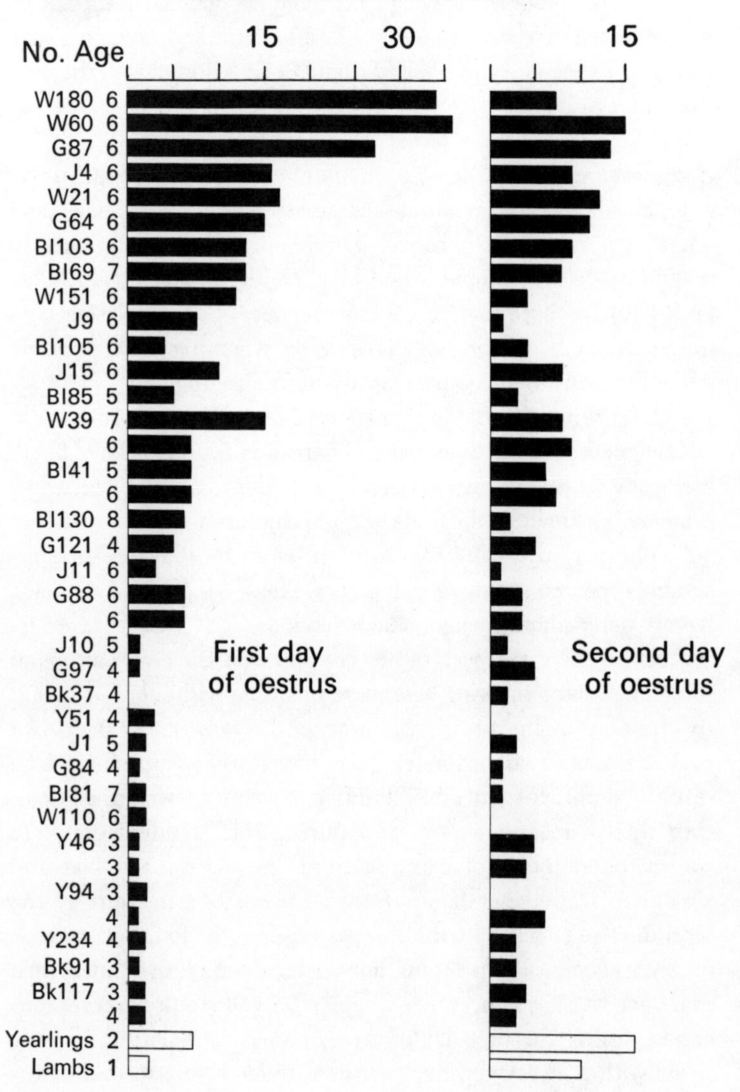

No. of oestrous ewes tended
No. Age
15
30
15
W180 6
W60 6
G87 6
J4 6
W21 6
G64 6
Bl103 6
Bl69 7
W151 6
J9 6
Bl105 6
J15 6
Bl85 5
W39 7
6
Bl41 5
6
Bl130 8
G121 4
J11 6
G88 6
6
J10 5
G97 4
Bk37 4
Y51 4
J1 5
G84 4
Bl81 7
W110 6
Y46 3
3
Y94 3
4
Y234 4
Bk91 3
Bk117 3
4
Yearlings 2
Lambs 1
First day of oestrus
Second day of oestrus

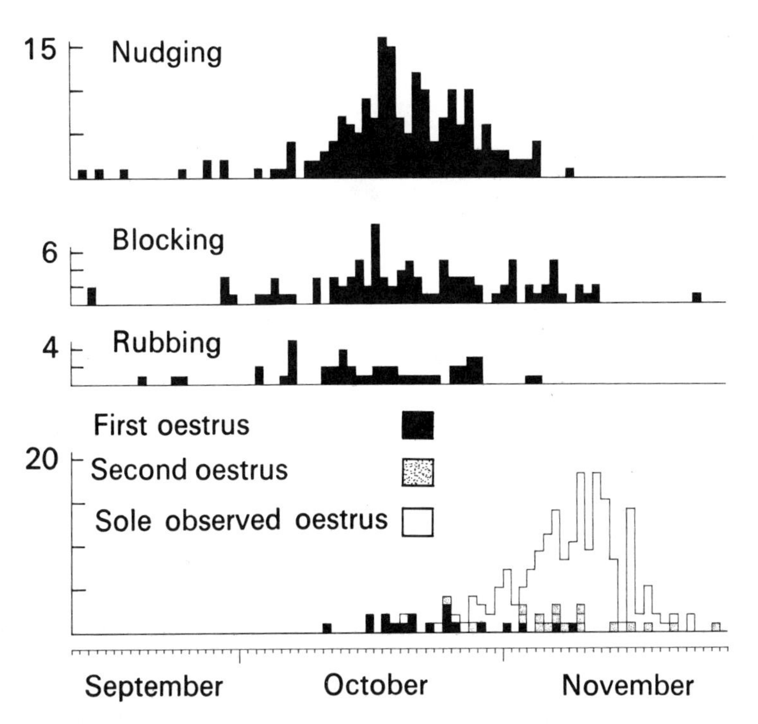

Fig. 3-9. Frequency of occurrence of behavioural interactions between Soay rams, as related to the incidence of oestrus amongst ewes in the study area on Hirta, St Kilda, 1966. (From P. Grubb and P. A. Jewell. *Journal of Reproduction and Fertility* Supplement 19, 493, Fig. 1 (1973).)

Fig. 3-8. The tending of Soay rams in the first and second days of oestrus (in the rut of 1966). Each ram is identified by his serial number (if marked), and age to the nearest year. The rams are set out in the order of the linear dominance hierarchy that was recorded amongst them from agonistic encounters. The master ram, W180, is at the top. Data for yearlings and for ram lambs are from all individuals observed. (From P. Grubb. In *Island Survivors*, p. 213, Fig. 8-14. Ed. P. A. Jewell, C. Milner and J. M. Boyd. London; Athlone Press (1974).)

Fig. 3-10. Soay rams in agonistic display during the rut. (From P. Grubb and P. A. Jewell. *Journal of Reproduction and Fertility* Supplement 19, 493, Fig. 4 (1973).)

tending bond as rams increase in rank; low-ranking animals may spend more time fruitlessly pursuing ewes during the waning period of their oestrus, since during the early fertile period they are kept away by dominant individuals. But if the reward of dominance is high, so is the price. In most mammals the mortality rates of males are somewhat higher than females; this difference is greatly accentuated in the Soay sheep, as shown in Fig. 3-11.

Waterbuck. It is interesting to compare the hierarchical system of the Soay sheep, where young males may attempt copulation and occasionally succeed, with a strictly territorial species like the waterbuck *Kobus defassa*, which was studied by Clive Spinage in Uganda. No male waterbuck below the age of 6 years obtains a territory (Fig. 3-12), although puberty is reached at the age of two. Thus the young males in this system are totally excluded

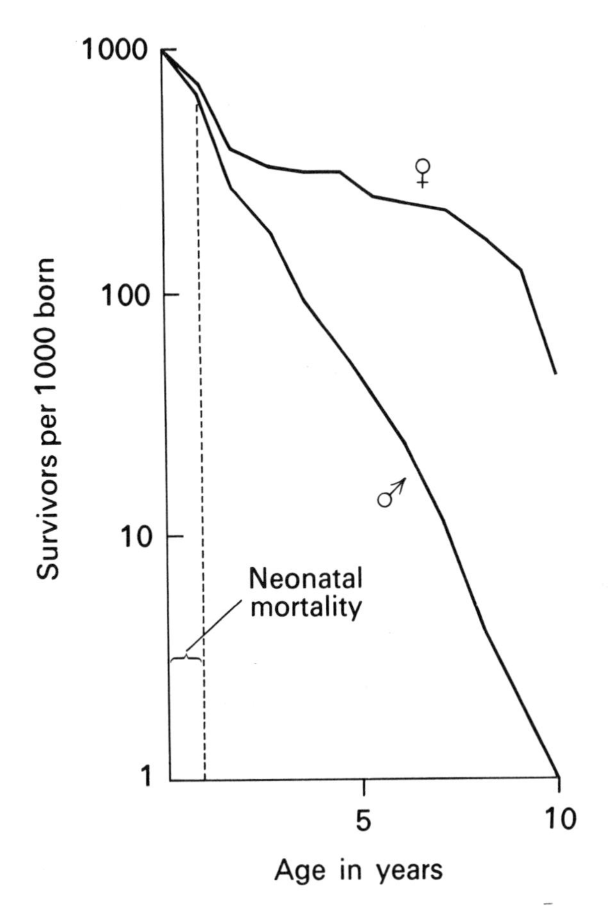

Fig. 3-11. Survivorship curves for Soay ewes and rams, Hirta, St Kilda. (From P. Grubb. In *Island Survivors*, p. 263, Fig. 10-10. Ed. P. A. Jewell, C. Milner and J. M. Boyd. London; Athlone Press (1974).)

from access to the females. The size of territory is related to body weight and horn growth, which both increase with age; thus the successful male increases his territory size year by year, up to a peak at about 9 years (Fig. 3-13). In favoured areas waterbuck males hold territories that abut on to one another and form a complete mosaic. Each territory provides an area that is adequate

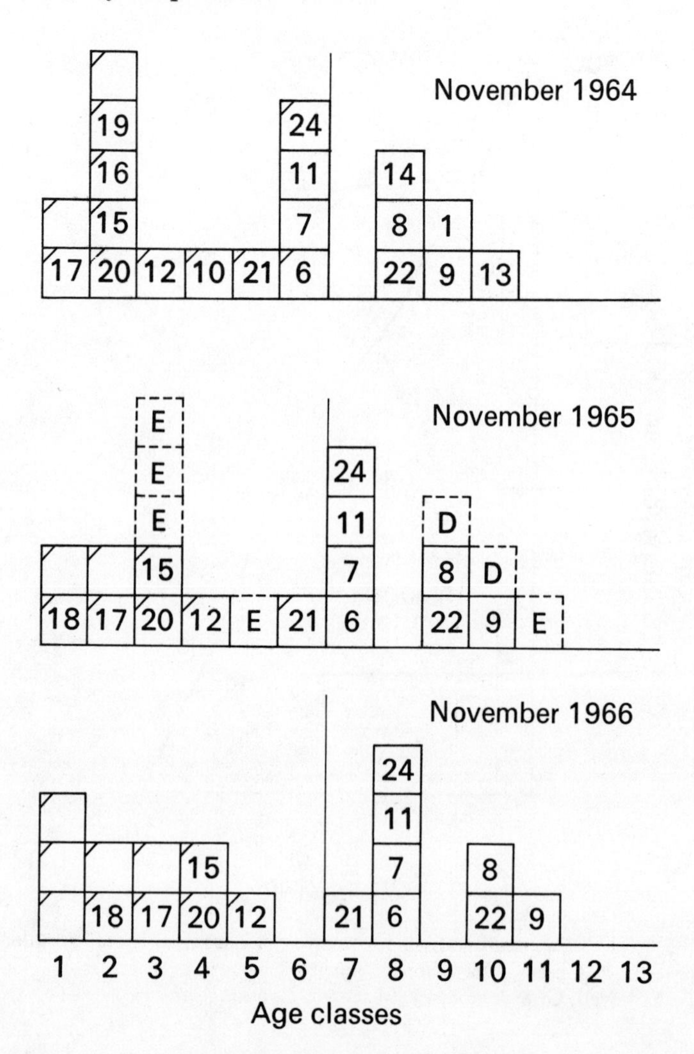

Fig. 3-12. Age structure of male waterbuck population on Mweya Peninsula, Uganda, in each of three successive years. Identification numbers of individuals are in the squares. Marked squares represent bachelor herd males that were not holding a territory, while open squares stand for territory holders. D, died; E, emigrated. (From C. A. Spinage. *Journal of Zoology, London* **159**, 339, Fig. 6 (1969).)

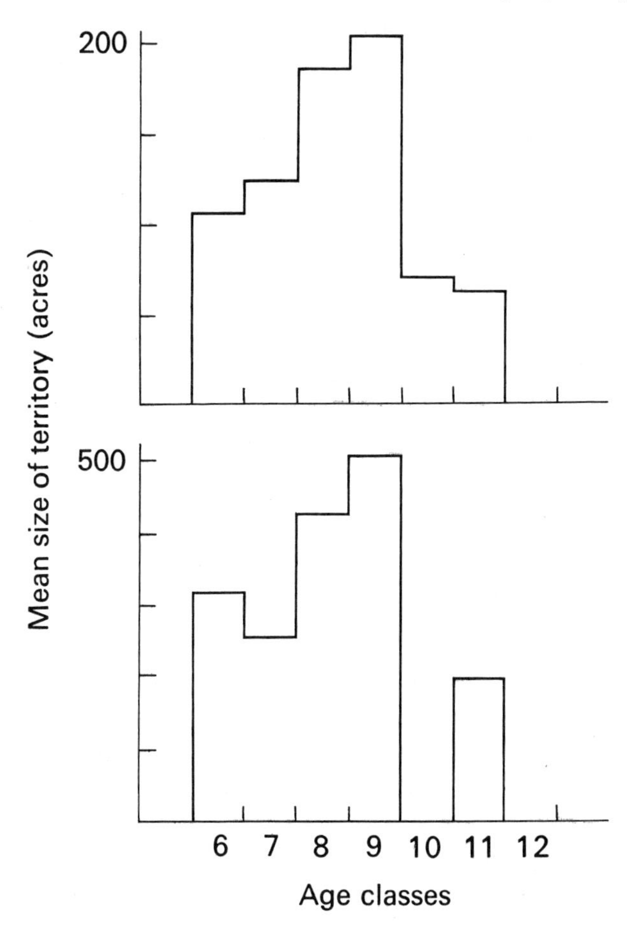

Fig. 3-13. Mean size of territory held by waterbuck males in each age class on two study areas in Uganda. From C. A. Spinage. *Journal of Zoology* **159**, 340, Fig. 7 (1969).)

for the male to feed, and the females wander freely from one territory to another.

Uganda Kob. The extreme form of territoriality that has been evolved by another East African antelope, the Uganda kob *Adenota kob*, presents a system that intensifies male competition.

Selection for reproductive success

At low population densities male kob may hold single, dispersed territories, but where the kob are exceptionally numerous, as in open grassland, territorial mating grounds develop. These are formed from a large number of individual territories that have become compacted together: there may be twenty or more, each being no more than 40 metres in diameter. A male stands 'at stud' on his territory, fiercely excluding all other males, but his stay is limited by sheer lack of food. The most favoured territories are those at the centre of the complex, and males in their prime fight vigorously for them. This situation reveals female choice in a striking way. Females in oestrus approach a territorial mating ground, and although they are first courted by the males at the edge, most females prefer to ignore these advances and move to the centre of the ground before accepting copulation. Thus it would appear that the females are selecting the males of highest proven fitness.

Hartebeest, topi and wildebeest. Three other African herbivores, the hartebeest *Alcelaphus buselaphus*, topi *Damaliscus korrigum* and wildebeest *Connochaetes taurinus*, all of the Tribe Alcelaphini, have evolved fascinating variations in their mating and calving strategies that reflect their respective adaptations to different habitats. All are grazers, and inhabit open grasslands, but the hartebeest shows a preference for bushy areas, whilst the wildebeest does not. Successful mature hartebeest males remain in fixed territories, but these have ill-defined areas of overlap at their boundaries, and the distances between individuals are important in determining the intensity of interactions. Similarly, both wildebeest and topi bulls establish static territories if the herds are permanently resident in one place; but in some parts of Africa both species can be migrating during the rut so that the bulls have to maintain their influence over the females whilst on the move.

I have studied the reproductive behaviour of topi at Ishasha in the Ruwenzori National Park in Uganda. This is a place where the animals live in a nomadic fashion. Their constant movement appears to create opportunities for a greater number of males to

Fig. 3-14. Two topi bulls, which are neighbouring herdmasters, engaged in a ritualized confrontation. They charge one another but drop to their knees just before impact.

take part in mating. The bulls engage in territorial behaviour, gather a group of females, and vigorously challenge neighbours (Fig. 3-14); it is only these herd-master bulls that mount and inseminate oestrous females. However, although the bulls maintain a sphere of influence analogous to a territory, their success is temporary, perhaps lasting only a day, because the females eventually break away and move on to fresh grazing. In so doing, the females become available to other bulls who capture them at the leading edge of a moving aggregation of animals. I have observed that success in capturing females by particular males can be related to specific places in the terrain through which the large herds of topi move. Recognizable males were successful as herd-masters at certain places, but elsewhere and at other times these same males were to be seen in 'bachelor' groups. Thus although there appears to be a fairly high turnover amongst the bulls, so that many get an opportunity of mating, there are nevertheless certain constraints on male reproductive success. Although puberty occurs in both sexes at 1½ years, males under the age of 3½ years were rarely seen to be successful but females often conceive

Fig. 3-15. A male topi killed by a lioness. More males than females fall prey to lions and it is probable that their preoccupation with competitive display towards other males, and associated lack of caution, makes them more vulnerable.

at puberty. Furthermore predation on males (Fig. 3-15) is heavier than on females, so that the sex ratio in the adult population is about forty males to sixty females.

In contrast to the female Uganda kob, female hartebeest, topi and wildebeest do not appear to make a choice of mate, but they do illustrate the advantages of conformity to species-pattern and one of their most interesting reproductive strategies relates to the timing of parturition to favour maximum calf survival.

Hartebeest drop their calves throughout the year, with peaks in February/March, just prior to the main rains, and in July/August. The female hartebeest gives birth in secrecy and keeps the calf lying away from her for some time in the covering of protective vegetation. Topi have a single calving season in August/September although there are some out-of-season calves. The strategy of the wildebeest is more extreme and they have a

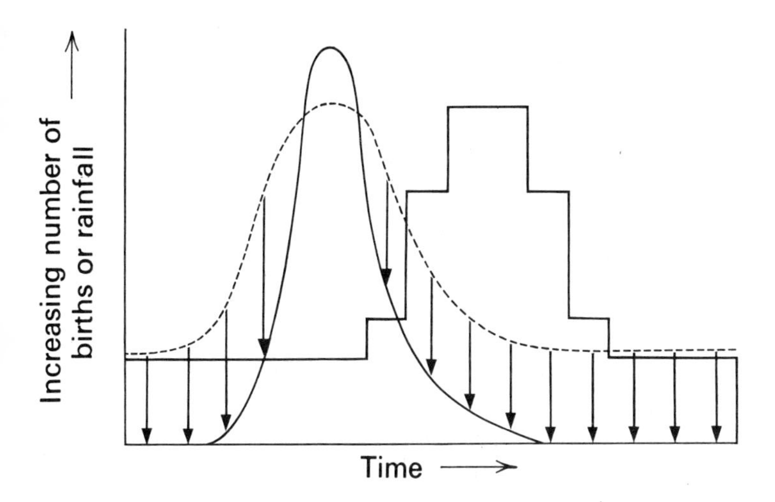

Fig. 3-16. Diagram to show how the distribution of parturition times (- - - -) which are primarily determined by a climatic condition (rainfall) (⎍) could be modified by predation (arrows) when this involves young born before or after the principal calving period. (From M. Gosling. *Journal of Reproduction and Fertility* 285, Fig. 12 (1969).)

highly synchronized calving season just before the usual occurrence of the rains.

Morris Gosling has suggested that this progressive concentration of calving time, as between the three species, has been evolved in response to the increasing predation pressure from lions and hyaenas that occurs in open country. A superabundance of calves born over a short period of time can swamp predator demand and thereby greatly increase the chances of an individual calf surviving. Conformity to the pattern is thus a strategy whereby the females achieve reproductive success. This concept is presented diagrammatically in Fig. 3-16.

WHAT IS THE ADVANTAGE OF ANTLERS TO MALE DEER

One further group of large herbivores that deserves mention is the Cervidae; in most species of deer the males develop an extraordinary adornment, the antler. Antlers are bony outgrowths from the skull, and their annual growth and shedding is controlled by the level of testosterone. This is such a remarkable process that it seems reasonable to conclude that the antlers must be of singular value to the male. The antler, when completely grown, becomes dead bone or 'hard horn' and Roger Short and his colleagues saw the opportunity this offered to experiment with the function of antlers – by cutting them off!

These experiments were carried out on wild deer *Cervus elaphus* on the island of Rhum in Scotland. The deer have a short rut in late September and October. There is intense competition between the stags at this time, and only the older and larger stags, which also usually carry the biggest antlers, successfully hold harems of females (Fig. 3-17). Two stags that had had their antlers cut off showed disrupted behaviour and failed to hold harems altogether, whereas they had been highly successful in the rut of the preceding year and were successful with their newly-grown antlers in the following year (Fig. 3-18). Stags with a congenital absence of antlers, however, so-called 'hummels', may occasionally be successful in holding a harem if they are only in competition with smaller, younger animals.

These results, showing the sexual significance of the antlers, also need to be seen against the complex nature of red deer

---

Fig. 3-17. The activity of red deer during the rut (phase 1, development of the rut; phase 2 (hatched), most active period of rut; phase 3, subsidence of the rut). The histogram shows the frequency with which hinds were seen with stags of different classes (*a*, 8–13 years; *b*, 5–8 years; *c*, 3–5 years; *d*, 1–2 years), or with no stag (*e*), in successive 5-day periods. The proportion of stags showing rutting behaviour is indicated by the graph. (From G. A. Lincoln and F. E. Guiness. *Journal of Reproduction and Fertility* Supplement 19, 479, Fig. 3 (1973).)

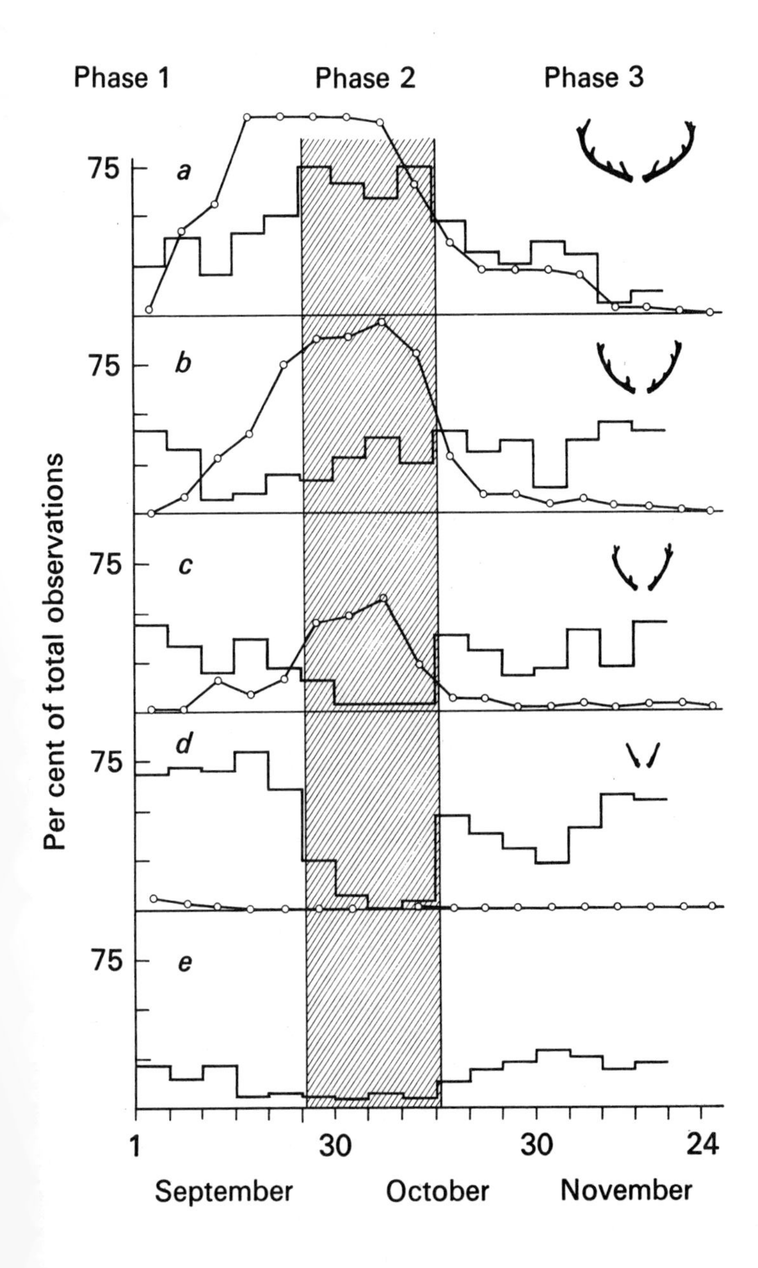

Phase 1
Phase 2
Phase 3
75
a
75
b
75
c
75
d
75
e
Per cent of total observations
1
30
30
24
September
October
November

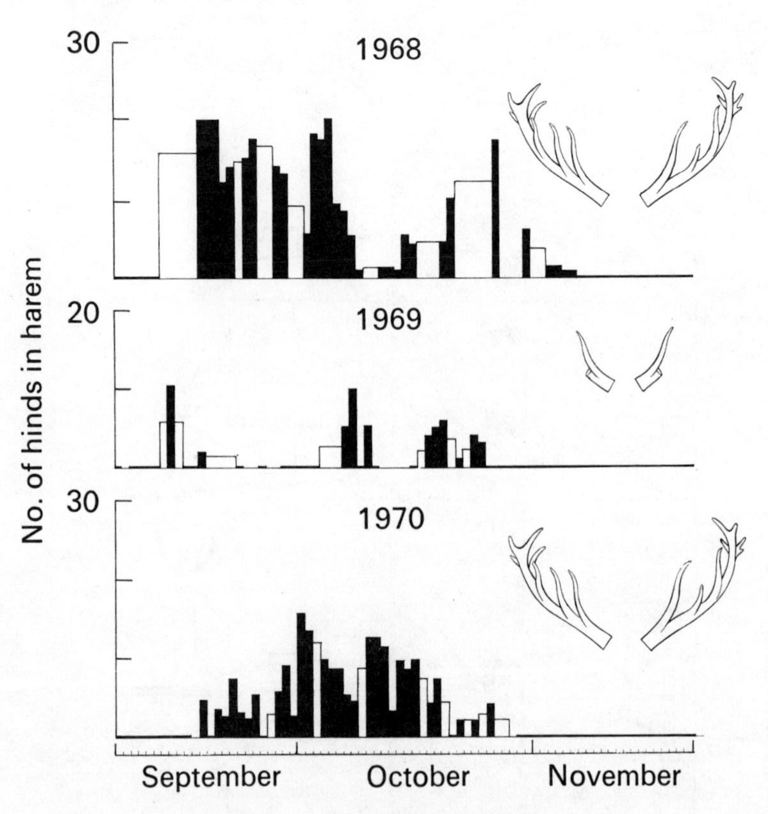

Fig. 3-18. Showing how amputating antlers affected the rutting success of a red deer stag, compared with its achievements in the years before and after the experiment. The histograms show the numbers of hinds in the harem of the stag day by day. Open columns represent the days when the stag was not seen. (After G. A. Lincoln. *Journal of Experimental Zoology* **182**, 233, Fig. 3 (1972).)

society. Out of the rut the stags live in bachelor groups, separately from the females. Within these groups, so long as the stags are in hard horn, a linear dominance hierarchy is firmly established. When the antlers are cast, and new ones 'in velvet' are growing, they are too sensitive to be used as weapons, so there is much temporary reshuffling of status. But when these new antlers harden off and lose their nerve supply with the shedding of the velvet, the weapons are restored and the old hierarchy is re-established. If the old antlers are sawn off a stag in a bachelor group, then the animal falls in rank. In an attempt to amplify the observations the research workers, with spectacular experimental flair, clamped an extra large pair of antlers on to a stag that until that time had a moderate head. This male had been number six in one of the hierarchies but, disappointingly perhaps, his enlarged antlers did not result in any increase of rank. Nevertheless, his re-introduction to the group, with his changed head adornment, greatly surprised, frightened, and provoked his comrades, proving that the antlers are also extremely important as individual recognition symbols.

Evidently the way in which red deer stags gain social dominance, and achieve ultimate reproductive success, is a complex matter. The antlers are a weapon, a shield, an organ of display, and a means of ready individual identification. There is a hint that the hinds are attracted to stags with large antlers during the rut, and if so this would further augment the exaggerated development of antlers by sexual selection. The antlers are not the only signal of strength, however, and large body size, dark colour, proud neck mane, and perhaps smell all play a part. Males that are able to form a harem one year are not necessarily successful in preceding or subsequent years. This is particularly likely to be so when a young male is first gaining dominant status, or when his powers are beginning to wane with old age, but fully adult red deer have been recorded as having this variability of success. Population density and abundance of males may well affect the situation. Deer surely offer great opportunities for further critical research.

MALES THAT KILL THE OFFSPRING OF THEIR FORMER RIVALS

As a final example of behaviour determining reproductive success in mammals, some recent work on lions in the Serengeti by George Schaller and Brian Bertram is of special interest.

Selection for reproductive success

Contrary to what may be generally supposed, lions do not always find it easy to obtain enough food. In East Africa wildebeest and topi (see Fig. 3-15) are preferred prey, but when these herbivores migrate in the dry season, the sedentary prides can go through periods of great privation and cub survival may be very low. The stable element of a pride is a group of lionesses often related to one another matrilineally. Males take charge of a pride – one, or two or more males acting together – but after a time the pride may be taken over by a rival group of males.

When a strange lion or group of lions succeed in taking over a pride they will sometimes kill the youngest cubs. Following the loss of their cubs, the mothers may come back into oestrus quite quickly (Fig. 3-19). Two immediate benefits follow for the male: he has eliminated some of the offspring of his supplanted rivals that would have competed with his own cubs, and he has speeded the return to oestrus and hence the availability for reproduction of at least some of his lionesses. It is also possible that he may have enhanced the chances of survival of his own offspring, because his action in killing cubs could help to synchronize a return to oestrus in the females. This would result in many cubs being born at about the same time. Their chances of survival would be higher than those born sporadically because, as all the lionesses would be lactating, the cubs could be suckled by several females, and there would be fewer older cubs to harass them. Furthermore, when male cubs from one of these synchronous peer groups eventually mature and leave the pride, they form friendly pairs or threesomes, and, through their combined strength, their chances of taking over a pride are higher than those of a single male. Similarly the young females, being of one age, are likely to come into oestrus close together and so perpetuate the phenomenon of synchrony. Brian Bertram has aptly summarized the advantages of this complex social organization: 'A pride that preys together, stays together'!

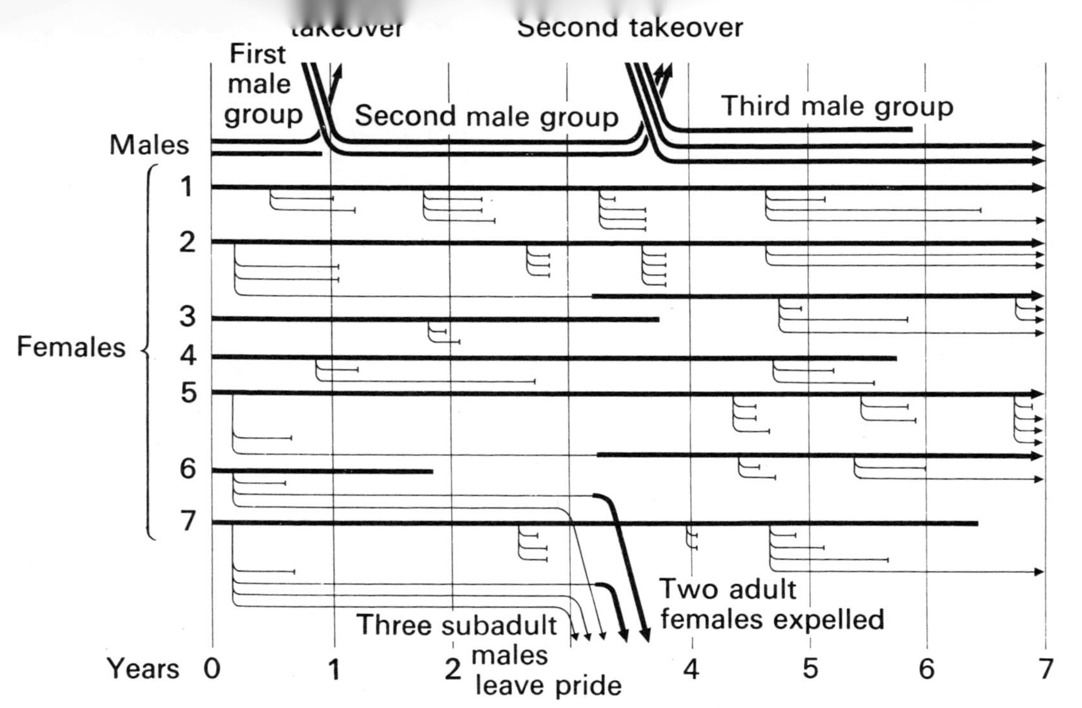

Fig. 3-19. A record of events in a pride of lions during 7 years of observation. The adult males in the group are changed twice through challenge from outside. Termination of lines in the diagram indicates the death of an animal; mortality among cubs was high, sometimes owing to slaughter by incoming males. (From B. C. R. Bertram. *Scientific American* **232**, 58 (1975).)

SOME THEORETICAL CONSIDERATIONS

The examples that have been used so far demonstrate how great the variance in reproductive success can be for the individuals in a population. For most species, male reproductive success is much more variable than female reproductive success. The ecology of a species, including its social organization and mating system, provides the context within which individuals must deploy their energies in order to maximize their chances of breeding and leaving viable offspring. This goal leads to intense intraspecific competition for a number of resources. One of these resources is the opposite sex, and sexual selection is a necessary outcome of competition for reproductive success. The variation in reproductive success, and the effects of sexual selection, will be markedly different in monogamous and polygamous species.

In monogamous species, male reproductive success would be expected to vary in the same manner as that of the female. The very fact that monogamy is the preferred mating system implies that the conditions for successfully rearing offspring are tipped in favour of co-operation between the two parents, and the male achieves greater reproductive success by resigning himself to this. Even so, male reproductive success may vary more than that of the female. The female, restrained by pregnancy and lactation, has nothing to gain from defying the system, but the male, though promiscuity, might gain a few bonus offspring.

The conditions that must be fulfilled for polygamy to operate successfully may be extremely exacting, and individuals must maximize their chances by an appropriate investment of time and effort, and also discernment in accepting a mate. These variables have been analysed in a number of theoretical studies.

Mate selection, parental investment and the investment of time in reproductive activities

For the female mammal, the recognition of a highly 'fit' male is of great importance. Once she has conceived she will be committed to a long pregnancy, and much effort and energy may be required in caring for her baby. She must reject unsuitable mates, and this could impose on the males the need for more appealing courtship behaviour. In most mammals, the female does not seek to copulate frequently. Indeed, in species where the duration of female receptivity during oestrus has been studied intensively, as in sheep, copulation has been shown to shorten the duration of heat.

The male mammal invests very little effort in any single act of copulation, and is usually capable of frequent ejaculation. However, he may have to engage in time-consuming courtship, or consortship, in order to be in possession of the female when she comes into oestrus. This leads to investments of effort that have been called 'courtship persistence' and 'female-guarding'. There is a danger that his time will be wasted if the female concerned should prove to be a poor mother. In mating systems where the male is dominant, or holds a territory, and therefore has undisputed access to a number of females, copulation with some unproductive females will be of little consequence, and the 'waste' of effort by the male will be insignificant. This is an obvious impetus for the evolution of dominance hierarchies and territory-holding systems.

One problem that a male has to face in developing an optimal reproductive strategy is how to apportion the time available for reproductive activities between staying with one female or searching for others, in order to achieve the largest number of pregnancies. Similarly the extent of parental investment in an offspring, necessary to ensure good chances of survival to reproduce its kind, must be weighed against the time the male might otherwise have spent in searching for other females. Since the total number of offspring produced by one sex in a population cannot exceed the total number produced by the other, it follows that the sex whose parental investment is the greatest, and in mammals this is usually the female, will become a limiting resource for the other sex.

Parental investment by the male mammal may take many forms, and the very low parental investment made by many male herbivores, for example, must not be over-emphasized. The male may defend the female, as in lions or certain bovids, and may provide food, as in hunting dogs and wolves. The male may also provide social status for his own young, with consequent benefits, and he may also provide a more general benefit like group protection or transmission of learned behaviour, as in many primates. Such male activities all tend to decrease the disparity in parental investment as between males and females.

The males of some species may be obliged by other social or environmental factors to accept a relatively small share of the female 'resource'. For example, the constant wide-ranging food-searching forays of female topi during the rut provide opportunities for a large number of males to participate in mating. The highly synchronized incidence of oestrus in the female wildebeest of the Serengeti plains, which as we have seen is ultimately a protective mechanism for the newborn calves, must also increase the opportunities for the less dominant or less experienced males to impregnate females.

Time investment by the male may not even be restricted to the immediate proximity of mating. For example, I have observed that in topi the males maintain their female herding behaviour throughout the year although mating itself is seasonal. And in Soay sheep, young ram lambs only 6 months of age attempt to participate in the rut although they achieve no mating success against more experienced rams. However, this does enable them to exercise and refine their reproductive behaviour, which is a valuable dividend from the time they have invested.

Sexual selection

All these patterns of behaviour we have described lead directly to sexual selection; this is obvious in the case of the male. But there is also some degree of sexual selection amongst female mammals, although the modes of competition are usually less obvious. Dominance hierarchies amongst females function in this way; furthermore, dominant males may accentuate the process by preferring to consort with dominant females, as in the case of some primates.

No reference has been made so far to the manner in which selection for reproductive success may have worked or be working in our own species, although this topic is discussed briefly by Roger Short in the next chapter. The subject is overlain in such a complex way by human social institutions that it would be impossible to do full justice to it here, but there are some points that should be made. In addition, I would like to put forward one idea concerning selection of infants, since this mechanism has not been implicated in any other mammalian species.

Darwin was much concerned with the way in which beauty in women might have been selected. For such selection to be operative, females well endowed with those traits would need to leave behind proportionally more infants that reached puberty than less beautiful women. But regardless of whether our ancestors were polygynous or monogamous, it is difficult to imagine that any fertile women, however ill-favoured, would have remained unmated or even suffered any reduction in fertility on this account. The promiscuous habits of men would have guaranteed their fertility! Darwin suggested that in polygynous families of primitive people, men of prestige would have selected attractive women. This, in itself, would only have led to an evolutionary advantage if other women mated to poorer men were less reproductively successful. This might have been the case if the mortality of children was higher amongst these families of lower status, for example because more food, or other vital resources, were available to the superior groups, and the others suffered deprivation. Sexual selection and survival selection would have had to reinforce one another for 'beauty' to have emerged by this means.

It can be stated that a strictly monogamous mating system, in populations where the numbers of males and females are equal, would not permit sexual selection to operate. For man, however, this need not be true, since competition and mate selection are not the only operative factors. Man has the opportunity to bestow advantages on particular individuals amongst his own offspring; a pretty child might attract the extra attention that would ensure its survival when others died at a time of food shortage. Such favours might also be reserved for the children of a particularly attractive woman; situations in which children shown this preference do survive at the expense of their siblings have been described for some present day South American communities. We need to discover much more about how natural selection has operated in our species if we are to interpret our physique and understand our behaviour.

Clearly a study of the behavioural factors that have promoted reproductive success provides a fascinating insight into the way in which natural selection works. Both sexes contribute to the reproductive efficiency of a species. They do so in varying ways but their ultimate contribution must be equal. We have had to draw our conclusions from field observations of animals in their natural state and we have usually not been able to put the hypotheses to the test. Nevertheless so much has been learned from field studies that we should soon be able to quantify the rewards and hazards that are offered by the different mating systems. This cannot be done in the laboratory, and is a challenge for tomorrow's field biologists.

SUGGESTED FURTHER READING

A model for the evolution of pinniped polygyny. G. A. Bartholomew. *Evolution* **24**, 546–59 (1970).

The social systems of lions. B. C. R. Bertram. *Scientific American* **232**, 54–65 (1975).

Sexual Selection and the Descent of Man. Ed. B. Campbell. London; Heinemann (1972). (Especially chapters by E. Mayr on 'Sexual Selection and Natural Selection'; R. L. Trivers on 'Parental Investment and Sexual Selection'; and J. H. Crook on 'Sexual Selection, Dimorphism, and Social Organisation'.)

The significance of the pair-bond in the Kittiwake. J. C. Coulson. *Proceedings of the XV International Ornithological Congress*, pp. 424–33. Leiden; E. J. Brill (1972).

Suggested further reading

The Descent of Man, and Selection in Relation to Sex, vol. II. C. Darwin. London; John Murray (1871).

The social organisation of mammals. J. F. Eisenberg. *Handbuch der Zoologie* **8**, 1–92 (1966).

Social organization and movements of topi (*Damaliscus korrigum*) during the rut, at Ishasha, Queen Elizabeth Park, Uganda. P. A. Jewell. *Zoologica Africana* **7**, 233–55 (1972).

Island Survivors: the Ecology of the Soay sheep of St. Kilda. P. A. Jewell, C. Milner and J. M. Boyd. London; Athlone Press (1974).

The social and sexual behaviour of the red deer stag. G. A. Lincoln, R. W. Youngson and R. V. Short. *Journal of Reproduction and Fertility* Supplement 11, 71–103 (1970).

4 The origin of species

R. V. Short

In the whole of biology there can be no concept that is at once so fascinating, so simple in outline and yet so baffling in detail, so much discussed and yet so little understood as the origin of species. Darwin's basic hypothesis of evolution through natural selection has stood the test of time, and continues to fire the imagination of individuals and provide the inspiration behind new scientific disciplines. The *Drosophila* geneticists have now produced experimental proof of the role of mutation and selection in evolution. Today, we live in the age of the molecular biologists, who have done so much to unravel the very nature of the bricks and mortar from which all living matter is constructed. But one cannot deduce the function of a building merely by a molecular analysis of its cement. Is it not time that some of us returned to the whole animal, and attempted to integrate our detailed knowledge of its component parts into an understanding of the ecology of the individual in its natural habitat?

Glancing through textbooks on evolution seldom reveals examples taken from amongst our surviving mammals, and this chapter is therefore a brief attempt to redress the balance by considering specifically the mammalian evidence about the ways in which species may have originated.

Although natural selection offers a simple and acceptable way of explaining how Nature exerts a constant pressure to adapt and improve *within* a species, the central problem is how do *new* species originate? It is easy enough to understand how naturally occurring variability between individuals is the substrate on which natural selection acts, but considerably more difficult to understand the mechanisms by which divergent selection ultimately leads to speciation. What ways has Nature used to prevent an advantageous mutation, occurring at random in a single individual at a single point in time, from becoming smothered by the adverse genetic load of the remainder of the population? Does each emergent species require its own Galapagos in order to prevent cohesion of its gene complex with that of the parent population from which it is struggling to escape? Or are there other ways of achieving reproductive isolation without having to go to the extreme lengths of geographical separation? Darwin puzzled over this very problem, and during his later years he put forward the concept of sexual selection, largely in order to explain the coexistence of so many different races of the one species, man. As Peter Jewell has discussed in the preceding chapter, we now tend to regard sexual selection (or assortive mating) as merely another form of natural selection. But there is always the lingering doubt as to whether extreme consequences of sexual selection, such as the massive antlers of the great Irish deer, might not eventually become disadvantageous and operate in opposition to natural selection (see Fig. 4-1).

Bateson, the geneticist, summarized the continuing dilemma about the origin of species when he said in 1922: 'That particular and essential bit of the theory of evolution which is concerned with the origin and nature of species remains utterly mysterious...the production of an indubitably sterile hybrid from completely fertile parents which have arisen under critical observation from a single common origin is the event for which we wait'. Perhaps we wait in vain, because this is probably not the way in which species are formed in the first place.

Bateson clearly envisaged speciation as occurring by means of a sudden, spontaneous genetic change within one homogeneous population, which there and then cleaved it irrevocably into two reproductively isolated groups kept apart from one another by hybrid sterility. But the essential weakness of this view of speciation is that it presupposes the *simultaneous* occurrence of the *same* rearrangement in a male *and* a female, who must then be able to meet and mate in order to propagate the newly emergent species; the odds against this happening must be long indeed. If genetic change, rather than geographical isolation, is ever to be the

Fig. 4-1. The great Irish deer *Megaceros giganteus*. Did sexual selection result in such massive antlers that the animal was ultimately doomed to extinction?

precipitating event in the formation of a new species, it would be better to postulate that some more subtle and less efficient reproductive isolating mechanism was present in the initial stages, allowing a single mutant individual to propagate that mutation within the population at large. No genetic difference can ever become established in a population without first passing through a stage of heterozygosity. Hybrid sterility should therefore be regarded as the end result and not the initial cause of the speciation process.

WHAT IS A SPECIES?

It is important to realize at the outset that terms such as species, sub-species, race and strain are man's attempts to divide what is essentially a continuum of genetic change into an arbitrary series of discrete stages. It is as unrealistic to imagine that a single genetic event can lead instantaneously to a new species as it is to accept the literal truth of the Bible's account of the Creation.

The old definition of species status was that individuals of different species were incapable of producing fertile hybrids when mated together. This distinction still holds true in extreme cases; horses, cows, sheep, pigs, dogs, cats, rats and mice will not produce hybrids at all, even when mated artificially with one another. But the definition begins to break down when one considers more closely related animals. Thus a domestic bull *Bos taurus* can be crossed with a North American buffalo cow *Bison bison* to produce a hybrid, the cattalo, which is fertile, albeit at a reduced level. The polar bear *Thalarctos maritimus* hybridizes readily in zoos with the brown bear *Ursus arctos* to produce fertile offspring. Zoos have also succeeded in crossing lions *Panthera leo* with tigers *Panthera tigris* to produce fertile ligers. The wapiti *Cervus canadensis* of North America and Eastern Asia, and the red deer *Cervus elaphus* of Europe have become feral in New Zealand, where fertile hybrids with a voice that is a cross between the wapiti's bugle and the stag's roar occur under natural conditions.

There are clearly sufficient examples of interspecific hybrid fertility to invalidate any simple definition of species status based on the dictum of hybrid sterility.

Ernst Mayr of Harvard, one of the leading contemporary exponents of evolutionary mechanisms, has defined a species simply as a population separated from others by a discontinuity. He lists three principal types of isolating mechanisms that could produce this discontinuity – geographical, ecological and reproductive – and concludes that in the final analysis it is only those individuals that are reproductively isolated from one another which can be considered as separate species. This would seem to be at variance with the facts quoted in the preceding paragraphs, although it must be admitted that all these exceptions to the interspecific hybrid sterility rule have been contrived by man. In the natural state, it is obviously true that different species which mix normally do not mate; if they did, and if the offspring were fertile and at no selective disadvantage, the two species would ultimately blend into one.

Taking all these facts into consideration, it would seem preferable to confer species status on a population whose discontinuity was naturally maintained by *any* type of isolating mechanism, whether it was geographical, behavioural or reproductive. This still leaves open to subjective value judgement the specific status of those isolates that show only minor phenotypic variation, e.g. in size or colour, from the main population. If left in geographical isolation for long enough, no doubt the bewildering variety of long-tailed field mice *Apodemus sylvaticus* that inhabit the isolated islands off the coast of Scotland would ultimately achieve true specific status. There is unlikely to be much gene flow between shire horses and Shetland ponies, although at present we regard both as belonging to the one species, *Equus caballus*. Ultimately, anything that acts to impair gene flow between two populations for long enough will favour speciation.

ISOLATING MECHANISMS

Ernst Mayr has pointed out that the most indispensable step in speciation is the acquisition of isolating mechanisms, and he believes that geographical isolation is the essential beginning for all new species; it is only *after* a group of animals has become separated by physical barriers that secondary isolating mechanisms begin to evolve. Whilst there can be no doubt that geographical isolation has been of overriding importance as the precipiating event in the formation of most new species, I should like to discuss in the pages that follow a few possible exceptions to this rule, where speciation may have occurred in the absence of geographical isolation.

But first, let us return to Ernst Mayr, and consider the array of secondary isolating mechanisms that effectively prevent cohesion of species even if the geographical barriers that initially separated them break down. Mayr has listed these in an orderly sequence under two main headings: premating mechanisms, that prevent the occurrence of interspecific fertilization, and postmating mechanisms that reduce the reproductive success of any interspecific crosses that may be formed. Since natural selection must operate to minimize reproductive wastage, it is tempting to visualize these isolating mechanisms as an evolutionary progression. For example, it is much less wasteful to prevent two species from even mating with one another, than to allow a pregnancy to go to term, only to result in the production of a sterile hybrid. Not only would such a hybrid effectively diminish its mother's reproductive potential by committing her to a long and fruitless gestation and lactation, but 'hybrid vigour' might allow it to compete successfully for limited resources, to the detriment of the fertile members of the herd.

Nevertheless, I think it would be wrong to assume that natural selection is necessarily operating to produce a less 'wasteful' isolating mechanism by shifting the reproductive block to an earlier and earlier stage of gestation. This presupposes that the two species must be in contact through a 'hybrid zone' where this selection could occur. There must be many more examples of species that have remained permanently geographically separated from one another, and yet have evolved isolating mechanisms entirely fortuitously, as a result of their increasing genetic divergence from one another. In Darwin's own words, 'Sterility is not a specially acquired or endowed quality, but is incidental to other acquired differences'.

Let us look at Mayr's list of isolating mechanisms, without necessarily regarding them as an evolutionary sequence, and consider some actual examples (Table 4-1).

TABLE 4-1. Interspecific isolating mechanisms

1 *Premating mechanisms that prevent interspecific crosses*
Potential mates do not meet (seasonal and habitat isolation)
Potential mates meet, but do not mate (ethological isolation)
Copulation attempted but no intromission (mechanical isolation)

2 *Postmating mechanisms that reduce reproductive success when interspecific crossing occurs*
Sperm transfer takes place but the egg is not fertilized (fertilization failure)
Egg is fertilized but hybrid dies early (early hybrid mortality)
F_1 hybrid dies later in gestation (late hybrid mortality)
F_1 hybrid is fully viable but partially or completely sterile (hybrid sterility)

(After E. Mayr. *Animal Species and Evolution.* Cambridge, Mass.; Harvard University Press (1963).)

Mates do not meet

The world abounds with species that are normally separated from one another by physical barriers. We have already mentioned the fact that red deer *Cervus elaphus* and wapiti *Cervus canadensis* will hybridize with one another if introduced into the same area, as has happened in New Zealand. In their natural habitats they are clearly still dependent on the existence of geographical barriers (Fig. 4-2) to keep them apart and maintain their independent specific status. Sika deer *Cervus nippon* have now become feral in many parts of Britain as a result of escapes from deer parks, and have started to hybridize with wild red deer in the Lake District. The three species are obviously very closely related to one another, differing principally in size; they have similar chromosomal karyotypes, and apparently produce fertile hybrids with one another.

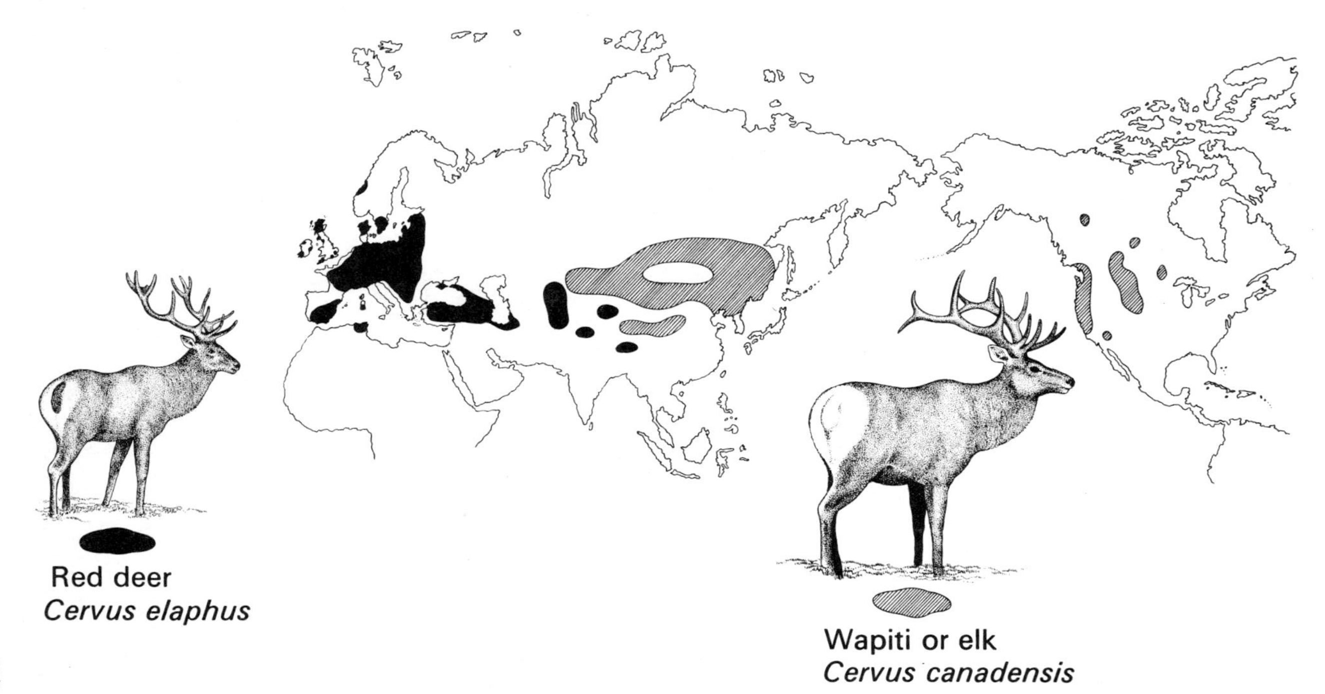

Fig. 4-2. Geographical distributions of the red deer *Cervus elaphus* and wapiti *Cervus canadensis*; under artificial conditions they will hybridize with one another to produce fertile offspring.

An intriguing example of seasonal isolation has recently been discovered in the spotted skunk *Spilogale putorius* of North America. Eastern forms mate in April and give birth to their young in June after a 55–65-day gestation period, whereas Western forms mate in September, and then there follows a period of about 200 days of delayed implantation, before implantation in early April and birth in May. So in contrast to the 2-month gestation of the Eastern form, the Western form has an 8-month gestation period. The timing of births in the two forms is broadly similar, but the widely differing mating times must serve as a most efficient reproductive isolating mechanism.

Potential mates meet, but do not mate

The Equidae are an unusual group of animals, since it is possible to produce viable though sterile interspecific hybrids between almost all the extant species, regardless of the major chromosomal differences that exist between them (Fig. 4-3). In the plains of Northern Kenya we find mixed herds of Grevy's zebra *Equus grevyi* and common zebra *Equus burchelli* but no hybrids have ever been reported, and no mixed matings ever seen. When one studies the social behaviour of the two species, as Hans and Utte Klingel have done, it becomes apparent that behavioural mechanisms have effectively established a complete reproductive barrier between them. The Grevy's zebra is territorial, with each male defending an area against intrusion by rival males, whereas the common zebras form stable family groups or harems of a stallion, his mares, and their offspring. Common zebras are not territorial in any sense, and once the harem has been established, a mare will always be covered by her own stallion. Evidently these non-territorial groups pose no threat to the highly territorial Grevy's zebra (once appropriately called the imperial zebra) who likewise does not threaten the sovereignty of the common zebra stallion over his harem.

Although the horse *Equus caballus* and the donkey *Equus asinus* originated in different continents, domestication has since brought them together, and man 'invented' those invaluable interspecific hybrids, the mule, and its reciprocal cross, the hinny. But it requires a great deal of ingenuity to persuade a jack donkey to serve a mare, or a stallion to serve a jenny; the signs of oestrus in the jenny and the mare are completely different, with the former revealing an 'assinine grin' and the latter merely 'winking'. Similarly, the jack's bray is a very different sound from the stallion's whinny. No wonder that mule breeders have to give the jack the sight of a jenny and may even need to play sweet music before he will serve a mare.

Copulation attempted, but no intromission

The most dramatic examples in this category occur amongst domesticated animals, where man has selected for extremes of body size. We have already referred to the improbability (impossibility?) of a shire stallion serving a Shetland mare, and similar situations obviously exist in the dog world, for example between the chihuahua and the wolfhound.

Fertilization fails

Although it is difficult to think of natural examples where mating occurs between species but fertilization fails to take place, this mechanism has been clearly demonstrated in laboratory situations. For example, Zeev Dickmann has shown that if rat eggs are surgically transferred to the oviducts of mated rabbits, rabbit spermatozoa will attach to the zona pellucida of the rat eggs, but will seldom if ever penetrate them. Yanagimachi has found that hamster sperm hardly ever penetrate the zona pellucida of rat or mouse eggs *in vitro*; even if the zona pellucida is removed, they still cannot penetrate the vitelline membrane of the denuded eggs. Guinea pig sperm likewise cannot penetrate the zona of hamster eggs; however, if the zona is removed, they are perfectly capable of penetrating the vitelline membrane and fertilizing the denuded egg. Yanagimachi has recently been able to show that even human

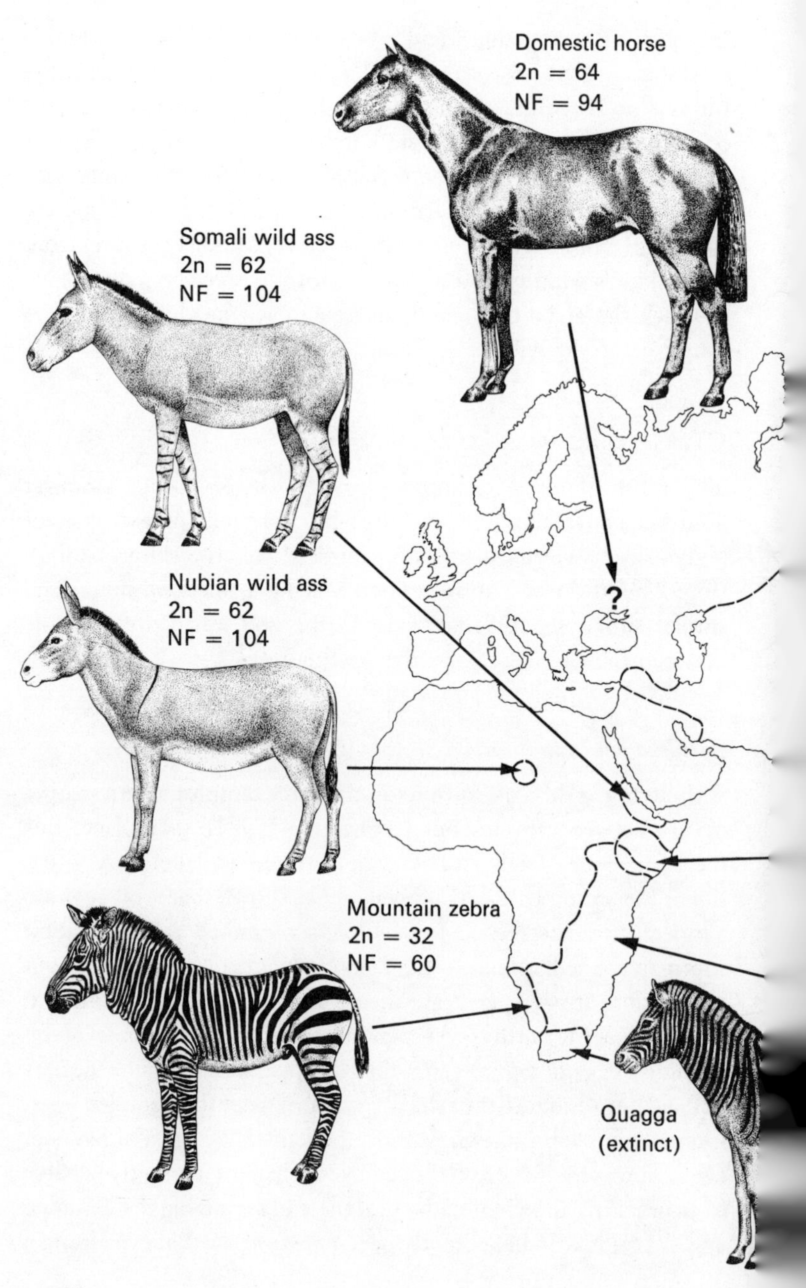

Fig. 4-3. Geographical distributions
number and Nombre Fondam

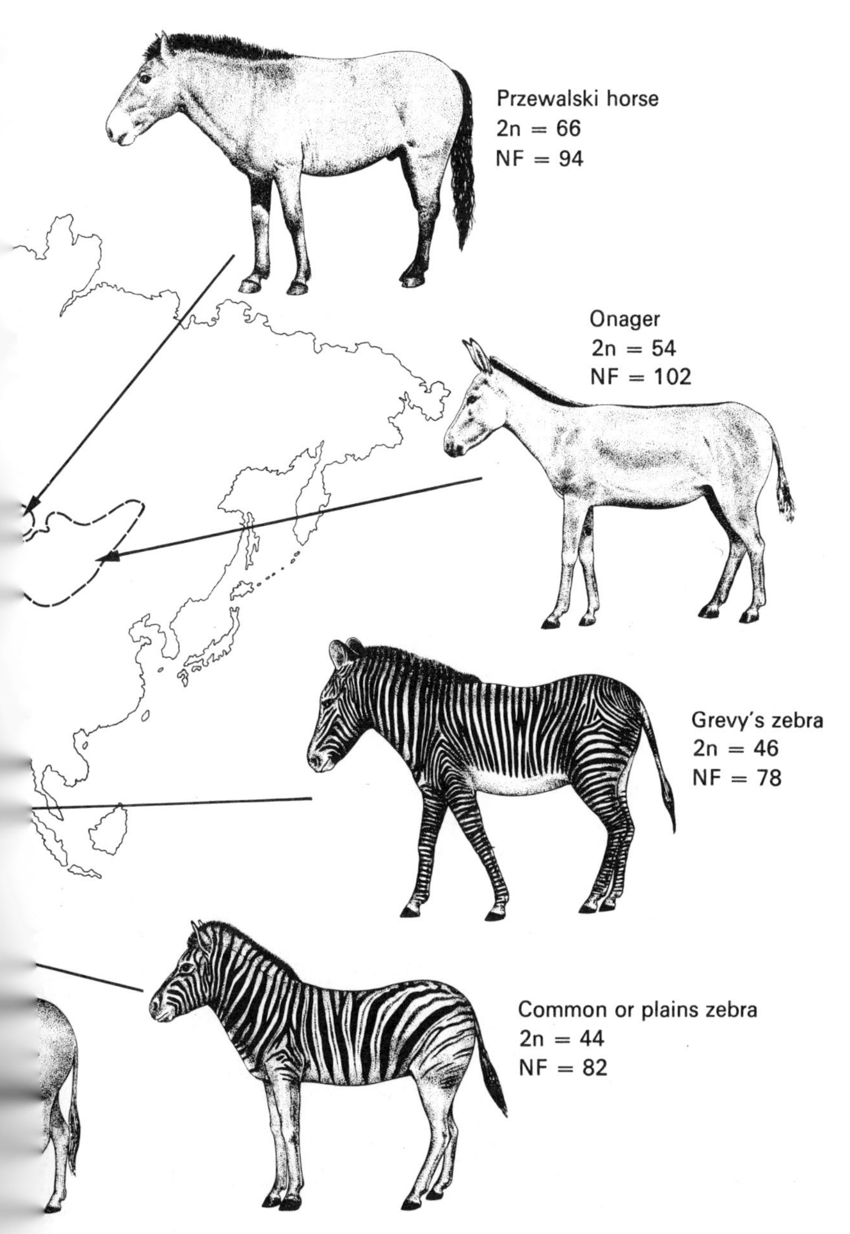

d Equidae, showing their chromosome
mber of chromosome arms).

spermatozoa can fertilize the denuded hamster egg (Fig. 4-4). These results emphasize the fact that there must be at least two separate barriers to interspecific fertilization, one located in the zona pellucida and one in the vitelline membrane.

Fertilization occurs, but hybrid dies early

Perhaps the best example of this situation is the 'rare habbit', first studied by Cyril Adams. If the domestic rabbit *Oryctolagus cuniculus* (2n = 44) is inseminated with the sperm of the common hare *Lepus europaeus* (2n = 48), fertilization appears to occur normally, and the eggs cleave to the late morula stage, when they all die (Fig. 4-5). M. C. Chang has produced similar crosses between the domestic rabbit and the cottontail rabbit *Sylvilagus transitionalis* or the snowshoe hare *Lepus americanus* (2n = 48), and he too found that although fertilization occurred, the embryo perished at the early blastocyst stage. Chromosome studies of these hybrid blastocysts showed that they had less than the expected modal chromosome number, the inference being that chromosome loss could be the cause of the early embryonic death.

F_1 hybrid dies later in gestation

In some interspecific crosses, fertilization, implantation and early embryonic development occur normally, but the fetus subsequently dies *in utero*. Perhaps the best example of this is seen in the case of sheep–goat hybrids (*Ovis aries*, 2n = 54 × *Capra hircus*, 2n = 60) which have been studied in detail by Jim Hancock and

---

Fig. 4-4. Hamster spermatozoa cannot penetrate the zona pellucida of a rat egg (*a*); if the zona is removed, they cannot penetrate the vitelline membrane (*b*). Guinea pig spermatozoa cannot penetrate the zona pellucida of a hamster egg (*c*); but if the zona is removed, they can fertilize the egg (*d*). Human spermatozoa can penetrate a hamster egg if the zona pellucida is removed and can initiate fertilization (*e*). These experiments point to the existence of at least two separate barriers to interspecific fertilization, one in the zona pellucida and one in the vitelline membrane.

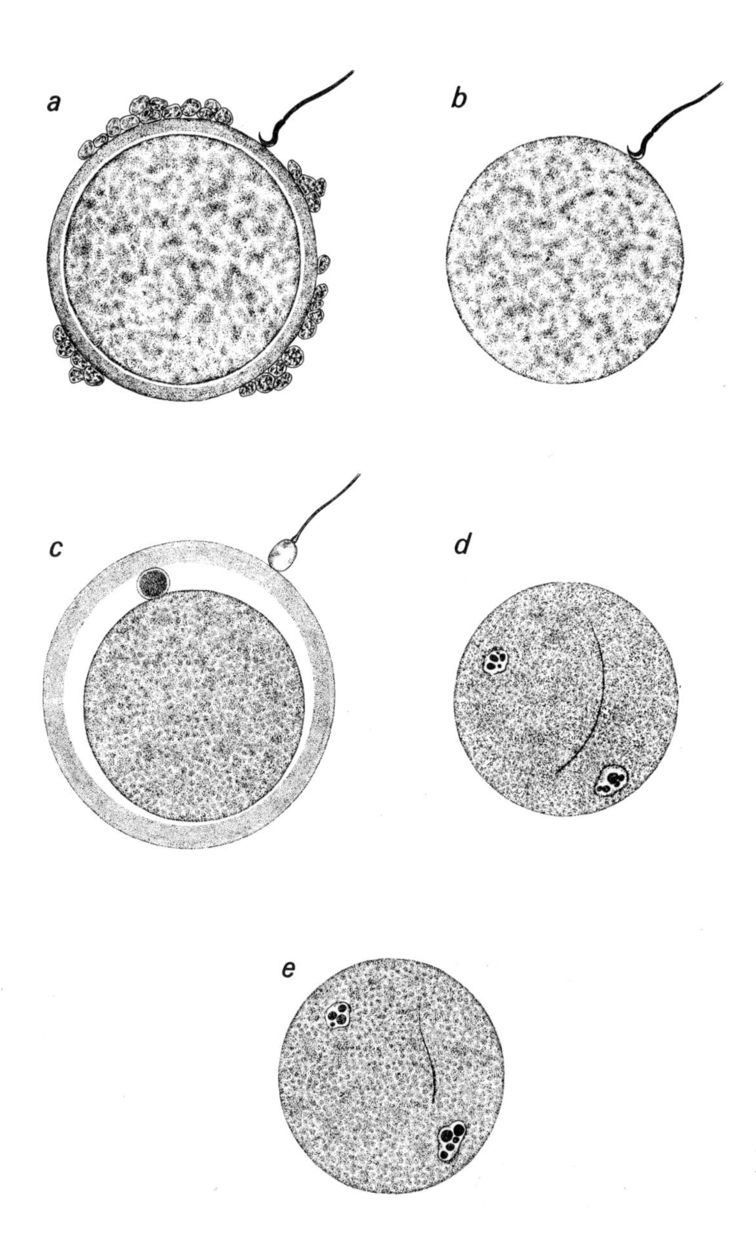
a
b
c
d
e

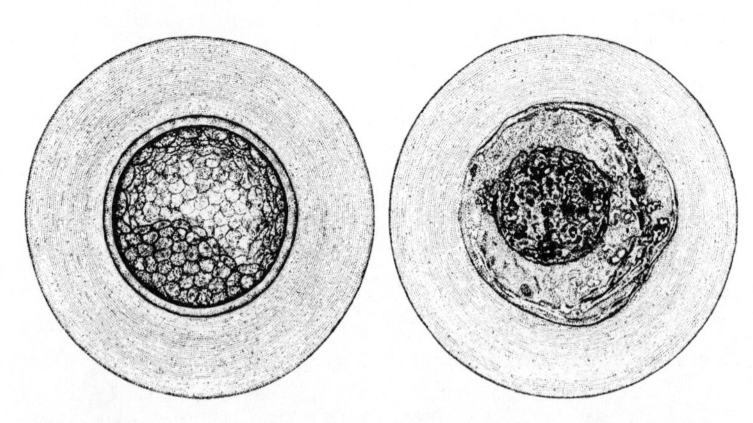

Fig. 4-5. A normal rabbit early blastocyst (*left*) and a degenerating hare ♂×rabbit ♀ embryo (*right*), which probably died because of missing chromosomes.

his colleagues. Ram sperm are much more successful in fertilizing goat eggs than vice versa, although we do not understand why. Once fertilization has occurred, the hybrid fetus develops quite normally in the uterus of the goat, and has the expected modal chromosome number intermediate between the two parental species (2n = 57). However, fetal death invariably occurs during the second month of gestation, apparently as a result of some immunological incompatibility between mother and fetus – a case of excessive maternal recognition of pregnancy? There could even be an endocrine mechanism for the pregnancy failure, because the sheep's placenta can produce progesterone from about the 50th day of gestation, whereas the goat's placenta never can.

Similar results have been obtained by M. C. Chang in ferret–mink hybrids (*Mustela furo*, 2n = 40 × *Mustela vison*, 2n = 30). Mink sperm are able to fertilize ferret eggs, but not vice versa. Hybrid embryos have the expected modal chromosome number of 2n = 35, and although initial embryonic development is normal, fetal death occurs at the end of the first month of gestation, again apparently as a result of some maternal rejection mechanism.

Hybrid is viable but sterile

The mule is the classical example both of hybrid vigour and of hybrid sterility. Although there is a comparatively small difference in chromosome number between the donkey *Equus asinus* ($2n = 62$) and the horse *Equus caballus* ($2n = 64$), this in fact conceals a greater difference in chromosome shape and size. If we count the number of chromosome arms (Nombre Fondamentale, NF) we find that the donkey has 104 and the horse 94, the donkey having 42 metacentric (X-shaped) chromosomes, and the horse only 30. The mule has the expected modal chromosome number of $2n = 63$, $NF = 99$, and, although the dissimilarities in size and shape of the two parental sets of chromosomes do not interfere with the normal process of mitotic cell division, the mule runs into difficulties in meiosis; pairing of homologous chromosomes prior to the first reduction division becomes virtually impossible (See Book 1, Chapter 2 and Book 4, Chapter 1). As a result of this, very few oocytes survive in the ovaries of female mules, so few follicles are formed, and the animals only occasionally come into oestrus. In the male mule, on the other hand, spermatogenesis is severely impaired, but Leydig cell function and libido are normal.

In spite of their meiotic block, female mules will occasionally ovulate, and male mules can produce a few abnormal spermatozoa (Fig. 4-6). It seems extremely unlikely that these gametes could be genetically balanced, so the possibility of a male or female mule ever being fertile must be remote indeed, in spite of occasional claims to the contrary.

It is surprising that there is no evidence of pregnancy failure, even in the more extreme interspecific equine hybrids, such as the zebra–horse and zebra–donkey crosses. Perhaps this has something to do with the rather primitive nature of the equine placenta, which is of the diffuse, epitheliochorial type; this means that it does not invade the maternal tissues, and is impermeable to maternal antibodies.

The foregoing examples have illustrated a variety of ways in

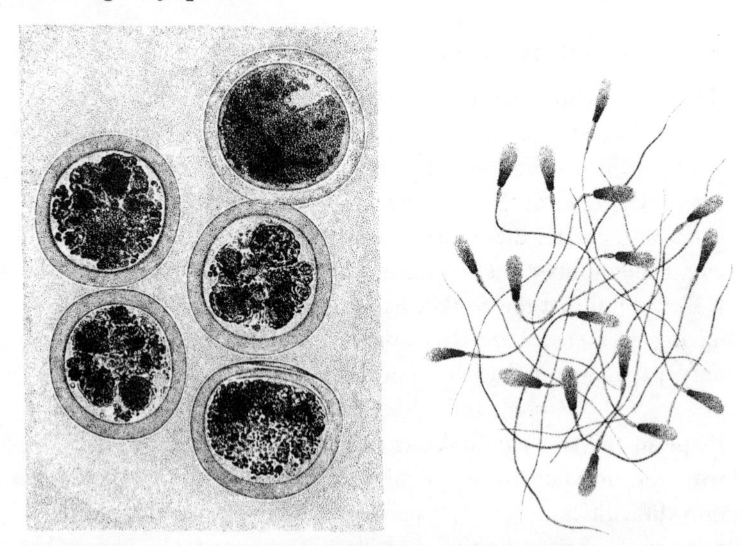

Fig. 4-6. Eggs recovered from the fallopian tube of a female hinny (horse ♂×donkey ♀) (left), and spermatozoa recovered from the epididymis of a male hinny (*right*). The chromosomal differences between the two parental species are so great that correct pairing of homologous chromosomes at meiosis in the hybrid's gonad becomes impossible, and it seems certain that these gametes must be genetically unbalanced.

which physiological mechanisms may operate to create a complete reproductive barrier between different species, thus preserving their genetic isolation. But it is important to re-emphasise the fact that these types of reproductive isolating mechanisms are probably an accidental *consequence*, rather than a cause of speciation. We must therefore return once more to the original question – how *do* species originate?

GEOGRAPHICAL ISOLATION

This has undoubtedly been the most powerful of all the speciating mechanisms, and if we devote comparatively little space to it, this is merely because its importance is already so well recognized.

The spectacular divergence of forms seen in our domesticated livestock gives some indication of the enormous latent genetic variability that must exist in all animals; differing selection pressures can produce a wide range of individuals in a comparatively short space of time. It is therefore not difficult to imagine how geographically separated members of a species could rapidly undergo a marked degree of divergent evolution if they were exposed to differing selection pressures, such as variations in habitat, food supply and climate.

But a response to different selection pressures cannot explain the surprising degree of morphological differences seen within species on neighbouring islands, where there are no apparent environmental differences. Darwin himself was obviously fascinated by this problem, so forcefully brought home to him by the great differences he observed in the birds, reptiles and plants on each of the islands in the Galapagos archipelago. In recent years, the zoology of archipelagos has been attracting increasing attention, and Sam Berry and Mike Delaney have made a particular study of the variability in the small mammals that inhabit the many islands around the coast of Britain. Quantitative measurements of differences in the skulls of long-tailed field mice *Apodemus sylvaticus* have enabled these workers to construct maps of the 'genetic distances' that separate populations on neighbouring islands. The general conclusion is that the races showing the *greatest* degree of divergence from the parental stock on the mainland are those founded by the *smallest* number of colonizing individuals; this highlights the great importance of the 'Founder Principle'. Given the extraordinary degree of genetic diversity that exists in all natural populations, it must follow that the smaller the sample you take, the less representative it can be of the population at large. Sam Berry concludes that this sampling artefact of the Founder Principle is the most powerful agent for gene frequency change. However, we must remember that it also carries with it certain inherent disadvantages for the isolated population, such as the danger of inbreeding homozygosity and consequent loss of fitness.

Another most important aspect of geographical isolation is the ecology of the species concerned. The more sedentary an animal, the more likely it is to become geographically isolated; the shorter its generation time, and the more young that are born, the more rapidly it will be able to respond to the selection pressures that lead to speciation. It can be no accident that the smallest and therefore least mobile mammals have the highest inherent fertility and show the largest number of species.

So far, we have considered the ways in which different types of isolating mechanism might allow speciation by favouring a subsequent change in gene frequency in the isolated population. But may be there are also ways in which a genetic change might itself become the isolating event.

CHROMOSOMAL CHANGE: A CONSEQUENCE OR A CAUSE OF SPECIATION?

As more and more species are examined cytogenetically, it becomes obvious that closely related individuals are sometimes separated by major karyotypic differences, whereas distantly related species may share apparently identical karyotypes. For example, the Chinese muntjac *Muntiacus reevesi* and the Indian muntjac *Muntiacus muntjak* are very similar in external appearance, but Kurt Benirschke has shown that *M. reevesi* has a diploid chromosome number of $2n = 46$, whereas *M. muntjak* amazingly has a diploid number of only $2n$ = ♂7, ♀6, the lowest so far recorded for any mammal (Fig. 4-7). At the other extreme, Ulfur Arnason has made a special study of the whales, and he has found that all the true mysticete or baleen whales, such as the Sei whale *Balaenoptera borealis*, have diploid chromosome numbers ($2n = 44$) and banded karyotypes identical to most of the Odontocetes or toothed whales, such as the common dolphin *Delphinus delphis* (Fig. 4-8). Hitherto, taxonomists had believed in a diphyletic origin of the whales; they presumed that the Odontocetes and the Mysticetes had evolved in parallel to one another from different ancestors. This new chromosomal evidence now makes it far more probable that both groups are derived from a common ancestor.

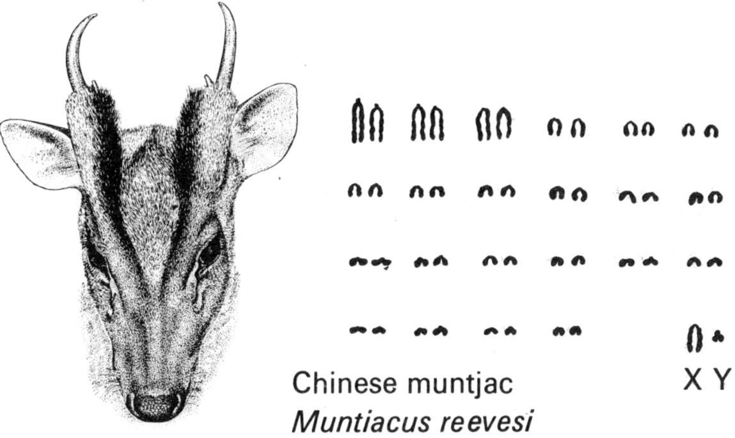

Fig. 4-7. The Chinese muntjac (*Muntiacus reevesi*) and its karyotype (*below*), and the Indian muntjac (*Muntiacus muntjak*) and its karyotype (*above*). The latter has the lowest known chromosome number of any mammal; the X chromosome appears to have become translocated onto an autosome. A similar $X/Y_1/Y_2$ sex-determining mechanism occurs in some marsupials (see Fig. 1-7). Is this spectacular chromosomal difference between two such closely related species a cause or a consequence of speciation?

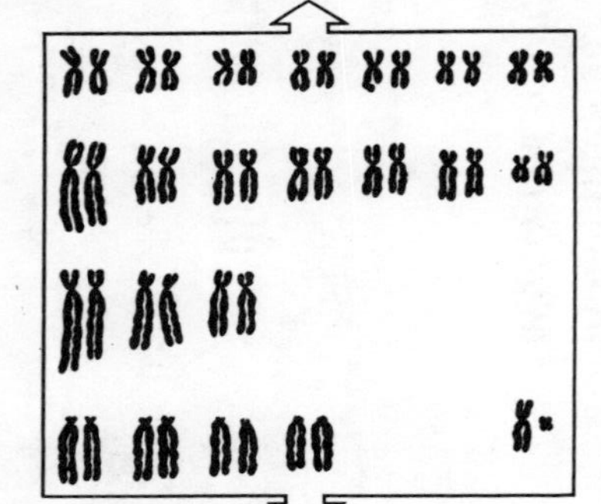

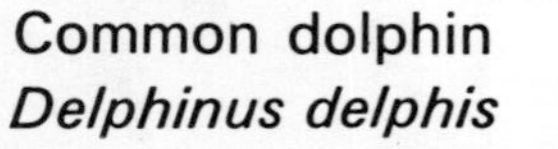

Fig. 4-8. The chromosomal karyotype of a mysticete, the Sei whale, is almost identical to that of an odontocete, the common dolphin. This suggests that the baleen whales and the toothed whales are derived from a common ancestor.

Whilst the whales provide further confirmation of the fact that speciation must often occur in the absence of any gross morphological changes in the chromosomes, the muntjacs make one wonder whether chromosomal changes might not occasionally initiate speciation. Michael White in Australia has championed this view, and has suggested that impaired hybrid fertility consequent upon chromosomal change might provide enough of a reproductive barrier to allow a new species to evolve whilst continuing to share a common habitat with the original population. White refers to this as 'stasipatric speciation', to distinguish it from the 'allopatric speciation' (geographical isolation) which Mayr regards as the only possible evolutionary mechanism.

Many types of structural chromosome change are possible, and in Book 1, Chapter 2 we discussed the meiotic complications that result from cross-overs, inversions and deletions. But one of the most frequent types of change that often seems to distinguish closely related species is the Robertsonian translocation. In this process, two V-shaped acrocentric chromosomes fuse at their centromeres to give rise to a single new X-shaped metacentric chromosome. Although this reduces the overall chromosome number by one, it does not change the NF (Fig. 4-9).

It can be appreciated that Robertsonian translocation of this type will not interfere with normal mitotic cell divisions, but it could give rise to problems at meiosis when homologous chromosomes have to pair in preparation for the first reduction division. If a Robertsonian translocation occurs in an animal, the one new metacentric chromosome will have to pair up with two acrocentrics, thus forming a trivalent configuration. When this trivalent breaks up at metaphase I, it is essential for the metacentric chromosome to go to one pole, and the two acrocentrics to go to the other; if there is non-disjunction of the chromosomes, a genetic imbalance will occur that will probably result in non-viable gametes or embryos (Fig. 4-9). The distinct possibility of meiotic errors occurring in individuals that are heterozygous for a Robertsonian translocation may provide the vital clue for White's stasipatric speciation mechanism. Let us see how it could work in practice.

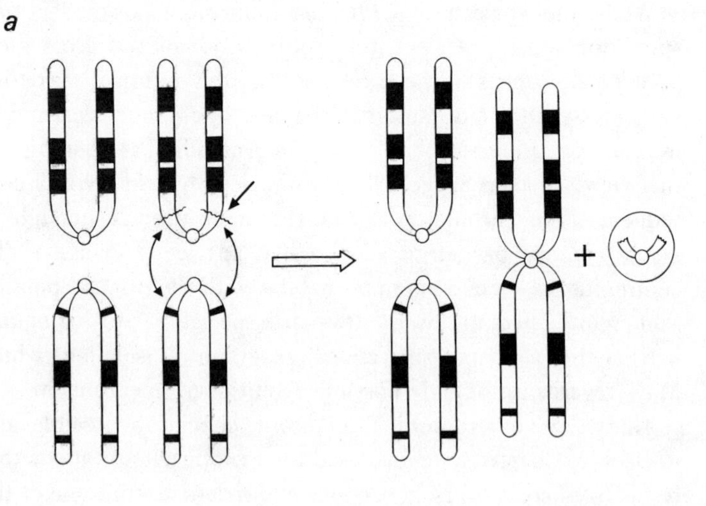

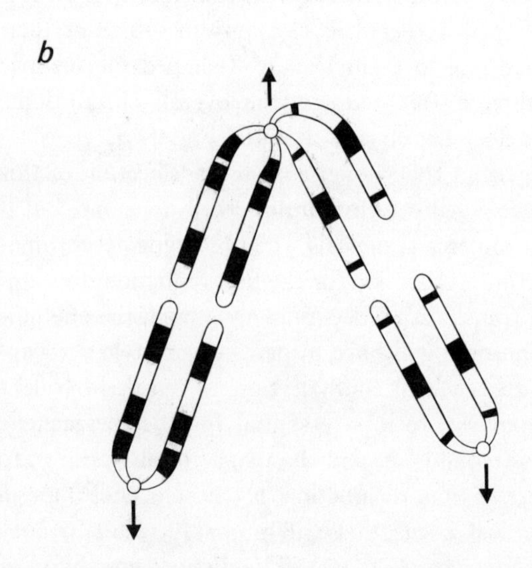

Fig. 4-9. *a* Robertsonian translocation, and *b* the complications that result when homologous chromosomes pair in preparation for the first reduction division at meiosis. The banding patterns correspond to those seen in the Przewalski horse/domestic horse translocation.

The evolution of the horse

All living forms of the domestic horse *Equus caballus* have 64 chromosomes, but no wild 64 chromosome equid exists (see Fig. 4-3). The nearest wild relative of the horse is the Przewalski horse from Mongolia, and this has 66 chromosomes (Fig. 4-10). Our own studies of the banded karyotypes of these two species have shown that a single Robertsonian translocation has occurred, four of the acrocentrics of the Przewalski horse being represented by two metacentrics in the domestic horse. When these two species are crossed, the hybrid has the anticipated modal chromosome number of 65 and is fertile; as expected, a trivalent is formed at meiosis.

Since there is only one proven mammalian example of fission of a metacentric to form two acrocentric chromosomes, it seems probable that the domestic horse could have arisen from the Przewalski, rather than the other way round. We can therefore construct a scenario to show how this could have taken place. Imagine a situation many thousands of years ago when herds of Przewalski horses roamed the vast plains of the Gobi desert. Two acrocentric chromosomes in one germ cell of one particular animal underwent a Robertsonian translocation to give rise to a single new metacentric chromosome, thus reducing the chromosome number in that cell from 66 to 65. This mutant cell would then produce a clone of germ cells with only 65 chromosomes, which would in turn produce gametes with 32 or 33 chromosomes. Therefore when this animal mated back to its parental 66 chromosome stock, up to 50 per cent of the offspring could exhibit the 65 chromosome mutation. In this way, the initial genetic event occurring in a single cell in a single individual could gradually be propagated through the population. There would eventually come a time when two 65 chromosome individuals might meet and mate with one another; such a union would have at least a one in four chance of producing a new 64 chromosome individual, with two metacentric chromosomes in place of the four acrocentrics of the Przewalski type. But if this chromosomal change was

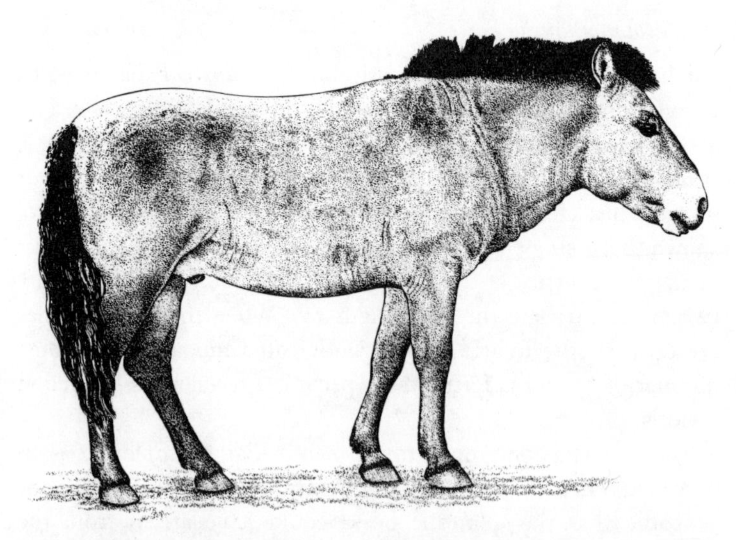

Fig. 4-10. The Przewalski horse, which has two more chromosomes than the domestic horse. First discovered in the Gobi desert at the end of the last century, and now probably extinct in the wild, it could be the ancestor of the domestic horse.

to give rise to a new 64 chromosome population capable of becoming a new species, and not just a balanced polymorphism of 64, 65 and 66 chromosome individuals in the one homogeneous population, the 64 and 66 chromosome animals would have to be at a selective advantage with respect to the 65. This is not improbable, since the trivalent formed at meiosis in the 65 chromosome animal might lead to a slight impairment in fertility, with the occasional chromosomal non-disjunction leading to the production of some genetically unbalanced gametes. It has already been shown in *Drosophila* by John Thoday that 'disruptive selection', or the simultaneous selection for two different characteristics within one population, can give rise to two divergent sub-populations which may eventually become genetically isolated from one another.

If Robertsonian translocations could have been of some evolutionary significance, it is interesting to consider whether they would have had more genetic impact if they had occurred in the male or the female. As Peter Jewell has pointed out in the preceding chapter, most mammals are polygynous, so that the successful male begets more offspring than the successful female. Furthermore, in the male, mitotic divisions of the germ cell line occur throughout adult life, so a Robertsonian translocation could occur at any time; in the female, mitotic divisions cease before birth (see Book 1, Chapter 2), so the opportunities for such a translocation occurring in a single germ cell to be propagated throughout the ovary are much more restricted. However, one must remember that a far higher proportion of the germ cells in the female are likely to give rise to offspring than in the male.

Separating the sheep from the goats

The domestic goat *Capra hircus* has 60 chromosomes, all of which are acrocentric, whereas the sheep *Ovis aries* has 54, of which three pairs are metacentric. When these karyotypes are studied in more detail using Giemsa banding techniques, the similarities are more striking than the differences. Both species have the same NF, and there are obvious homologies between the three metacentric chromosomes of *Ovis* and six of the acrocentric chromosomes of *Capra*. Since Robertsonian fusion of acrocentrics is thought to have been a commoner process than centromeric fission of metacentrics, it seems probable that the sheep karyotype has been derived from the goat type by three Robertsonian fusions.

The domestic sheep is probably derived from the wild mouflon *Ovis musimon* of the Mediterranean basin. But there are many other wild species of sheep that extend across Asia into North America, and the recent examination of their chromosomes has begun to unfold a fascinating story (see Table 4-2 and Fig. 4-11). The chromosome numbers vary from 52 (the snow sheep of East Siberia) to 58 (the urial sheep of Asia) and these differences appear to be due to variations in the number of Robertsonian fusions that have occurred. The domestic sheep will cross readily

Fig. 4-11. Geographical distribution of the diff
chromosome number that appears to r

breeds of sheep, showing the variability of
a series of Robertsonian translocations.

TABLE 4-2. The chromosome numbers of domestic and wild sheep and related species

Systematic name	Trivial name	Locality	NF	Chromosome no.
Ovis nivicola	Snow sheep	E. Siberia	60	52
Ovis aries	Domestic sheep	Europe	60	54
Ovis musimon	European mouflon	Europe	60	54
Ovis orientalis	Asiatic mouflon	Asia	60	54
Ovis dalli	Dall sheep	N. America	60	54
Ovis canadensis	Bighorn sheep	N. America	60	54
Ovis ammon	Arkhar or argali	Asia	60	56
Ovis vignei	Urial	Asia	60	58
Ammotragus lervia	Barbary sheep	N. Africa	60	58
Capra hircus	Domestic goat	Europe	60	60

with the mouflons, the argali, the urial and the North American bighorn, and apparently the hybrids are fertile, although sometimes at a reduced level.

But the situation is very different if one attempts to hybridize the domestic sheep with the Barbary sheep *Ammotragus lervia* or the goat. We have already seen how all sheep–goat hybrids die *in utero*. Barbary sheep spermatozoa can fertilize sheep eggs, but the blastocysts all die. This would seem to reinforce the taxonomists' contention that the Barbary sheep is more closely related to the goat than the sheep, a view that is further supported by the fact that the Barbary sheep will produce viable hybrids when crossed with goats. Presumably the Barbary sheep and the goat have been separated from *Ovis* stock for so long that secondary reproductive isolating mechanisms have already started to develop.

Recent studies of the chromosomes of wild sheep in Iran, where the ranges of Asiatic mouflon and the urial meet, have shown the existence of 54, 55, 56, 57 and 58 chromosome individuals. Are these polymorphic forms a result of the breakdown of geographical barriers that once separated the species, or are the species evolving from this polymorphic state as a result of the reduced meiotic efficiency of the heterozygotes?

From the evidence available, it seems likely that the bighorn and dall sheep of North America must have been derived from ancestors that crossed over the land bridge from Asia, leaving behind them a relict population that was to become the snow sheep of Siberia. Of course, it would be wrong to suppose that a Robertsonian-type chromosomal change by itself necessarily alters the genetic characteristics of the animal. Robertsonian translocations are occurring all the time in our domestic sheep, as Neil Bruere has recently shown in New Zealand, but so far there is nothing to distinguish the new 53 and 52 chromosome variants from the original 54 chromosome type, either in terms of phenotype or fertility.

The tobacco mouse

As long ago as 1869, Swiss biologists recorded the existence of an unusual looking, dark coloured mouse, inhabiting an old tobacco factory high up in the isolated Valle di Poschiavo in the Swiss Alps. This was recognized by taxonomists as being morphologically distinct from the house mouse *Mus musculus* and was given the name 'tobacco mouse' *Mus poschiavinus*. One hundred and one years were to elapse before Alfred Gropp and his colleagues at the University of Bonn examined the chromosomes of this new species, with surprising results. In contrast to the karyotype of the house mouse, $2n = 40$; $NF = 40$, with no metacentric chromosomes, the tobacco mouse has a diploid number of $2n = 26$; $NF = 40$, with seven pairs of metacentrics. When the tobacco mouse was crossed with the domestic mouse, a hybrid was produced with the expected modal number of chromosomes ($2n = 33$; $NF = 40$), and meiotic studies of the gonads of these hybrids showed the formation of seven trivalents, with one of the tobacco mouse metacentrics pairing with two house mouse acrocentrics in each case. This, together with banding studies of the chromosomes, leaves no doubt that the tobacco mouse karyotype is derived from that of the house mouse by seven pairs of Robertsonian translocations.

In contrast to the situation we have just described for the Przewalski horse hybrid, and some of the sheep hybrids, the fertility of these house mouse×tobacco mouse hybrids is seriously impaired; they only give birth to very small litters, and that infrequently. The infertility is a direct consequence of partial meiotic non-disjunction. In each of the seven trivalent configurations formed at meiosis, there is a high probability that one of the acrocentric chromosomes will fail to separate from its metacentric partner, thus producing a gamete that will give rise to either a trisomic embryo (possessing an extra autosome) or a monosomic embryo (lacking one autosome). A whole array of embryological defects can be produced, depending on which chromosome or chromosomes are involved in these trisomies or monosomies.

So how did the tobacco mouse arise in the first place? Do these translocations confer any special advantage on the tobacco mouse in its mountain fastness? It is interesting that there are now no true house mice to be found in the area at all. The degree of reproductive isolation already established between the two species is so great that it is difficult to imagine how any hybrids could survive, so we are presumably looking at mechanisms that have evolved since the tobacco mouse became geographically isolated from house mouse stock. If the Robertsonian translocations occurred one at a time, it is more probable that they could have survived as a balanced polymorphism, with gradual selection for the homozygote. No doubt the size of the original isolate was also a most important factor. Maybe a diet of tobacco helped to hasten the whole mutational process!

In trying to arrive at an informed view of the importance of chromosomal changes in speciation, all we can say is that the more we study the chromosomes of large populations of animals, the more we realize how widespread these balanced Robertsonian chromosomal polymorphisms are. First brought to light in the gerbil, and then demonstrated by Geoff Sharman, Charles Ford and John Hamerton in the common shrew *Sorex araneus* and since discovered in many species, including cattle, sheep and man, this naturally occurring chromosomal variability is an obvious substrate for natural selection. It is easy to see how the heterozygote could be potentially at a disadvantage as a result of decreased meiotic efficiency; any slight impairment of the reproductive success of the heterozygote would impede the gene flow between the two homozygous populations and favour speciation, even in a shared habitat. It is more difficult to explain why there is no apparent infertility in the naturally occurring translocation heterozygotes of domestic sheep, slight infertility in cattle heterozygotes, and extreme infertility in mouse heterozygotes.

In talking of chromosomal changes, we have only discussed Robertsonian fusions, but clearly these are by no means the only type of change taking place; if this were the case, all species would ultimately end up with a small number of chromosomes, all metacentric. The highest chromosome number in mammals is found in a South American fish-eating rodent *Anotomys leander* (2n = 92) closely followed by the black rhinoceros *Diceros bicornis* (2n = 84), whereas the marsupials universally have low chromosome counts, ranging from 10 to 32. However, the Indian muntjac, to which we have already referred, still holds the record for the lowest count in any mammal (2n = ♂7, ♀6). Robertsonian translocations occur relatively frequently and are easy to detect cytologically; hence they have received a disproportionate amount of attention. Centromeric fission appears to be a much less common event, although it has been described by Karl Fredga in the root vole *Microtus oeconomus* in Scandinavia.

WHENCE AND WHITHER MAN?

It would not be difficult, even for a schoolboy, to list an array of attributes that set man apart from all other animals: our large brain size, upright posture, and our spoken and written language, to name but a few. Undoubtedly these have been of major importance in our evolutionary history. The ability to teach and learn through the written and spoken word has allowed man to profit from the experience of previous generations, and 'inherit' technologies as well as genes. We have overtaken the blind, groping trial-and-error of Darwinian evolution, by our ability to transmit acquired knowledge. We will be forever creating new social and physical environments for ourselves, but since they are inevitably bound to be beyond our past evolutionary experience, many of these new situations will also be beyond our genetic competence. This may be our undoing.

Man has earned himself many epithets in recent popular writings: the tool-maker, the hunter, the meat-eater, the naked ape, the imperial animal. But there are many pitfalls for those who would attempt to reconstruct our past ecology from a mixture of fossil bones and speculation. An alternative approach has been to compare present-day man with his closest living relatives, the chimpanzee, gorilla and orang-utan. People still make the mistake of assuming that these anthropoid apes are primitive human beings, species that stopped evolving whilst man alone went on to greater heights. In point of fact, the chimpanzee is also highly evolved, and is supremely adapted for life in its tropical forest environment, far more so in fact than man is for life in a city.

Considering the millions of years that must have elapsed since the existence of a common ancestor for man and the great apes, one is more surprised by the similarities that exist between us than the differences. For example, our chromosomal karyotypes are very similar; all three great apes have 48 chromosomes, as compared to the 46 of man. Giemsa banding studies reveal obvious homologies between the chromosomes of the four species; the gorilla and man are also unique amongst mammals in having a Y-chromosome that fluoresces with quinacrine. Molecular biologists have recently used the technique of measuring differences in DNA base sequences between the proteins of different species to give an objective measure of the 'genetic distance' between them. Detailed studies of the proteins and nucleic acids of the human and chimpanzee show more than 99 per cent concordance; this gives an incredibly small genetic distance, comparable to that between sibling species of other genera. Since this genetic difference is manifestly too small to account for the major phenotypic differences that exist between the chimpanzee and man, this reinforces the view expressed by Susu Ohno in the opening chapter of this book that evolution has proceeded by mutations of regulatory genes that control the expression of characters, rather than by mutations of the structural genes themselves.

Social anthropologists have attempted to define the unique nature of man by studying the culture and customs of surviving primitive tribes. There can be little doubt that we have evolved from a primitive hunter–gatherer existence in our not-so-distant past, and the genetic attributes we possess today are a legacy of that life style. It is only eighty generations since the birth of Christ, and only a few hundred since the dawn of civilization, hardly time enough for any meaningful genetic change to have occurred. The life of hunter–gatherers today may therefore provide us with a mirror of our true selves. But once again, there are pitfalls; we cannot necessarily assume that present-day hunter–gatherers live similar lives to those of our primitive ancestors.

One of the things that sets us apart from all other primates is our sexual behaviour. Since there are no behavioural fossils, we cannot tell at what point in our evolution this change took place, and hence it must always remain a matter of speculation as to whether it is a cause or a consequence of speciation. But since reproductive success has always been the key to survival, we can perhaps tiptoe to some conclusions about the possible evolutionary significance of human sexual behavour.

As Peter Jewell has discussed in the preceding chapter, mammals that have adopted a polygynous mating system show a marked degree of sexual dimorphism; the males increase in size relative to females as a result of sexual selection. This rule seems to hold true for the primates also. The New World monkeys which are monogamous, such as the marmosets and tamarins, show little sexual dimorphism. The Old World monkeys are generally polygynous, with the females weighing at least 30 per cent less than the males; in the gorilla and orang-utan for example, the female is only half the weight of the male. The situation in the chimpanzee is somewhat different, since the female is only 10 per cent lighter than the male. This is probably related to the fact that chimpanzees live in mixed-sex troops containing several males, where there is generally little competition between males for an oestrous female, although temporary consort relationships may be formed between a female in oestrus and a particular male.

And what can we say about ourselves? Women are on average 20 per cent lighter than men, and there are obvious differences between the sexes in shoulder width, hip width, height, muscular strength and pitch of voice, to say nothing of the differences in cranial, facial and body hair that only become fully apparent in our unclad, unshaven and unshorn state. This degree of sexual dimorphism is in accord with our knowledge of present and past human social history; man seems to have strong polygynous tendencies. Although Christian cultures reject this way of life, Clellan Ford and Frank Beach carried out an anthropological survey of human sexual behaviour in 185 primitive human communities scattered throughout the world, and found that polygyny was the rule rather than the exception. However, this polygynous relationship often took the form of a succession of essentially monogamous long-term consortships, in other words, serial monogamy.

Long-term pairing has probably become necessary in order to provide us with a secure social environment for the rearing of our children, who take so long to grow up. The increase in brain size is one of the spectacular features of hominoid evolution, and since a large head is incompatible with birth through a narrow pelvis, much of the development of the brain has had to be deferred until after birth. This long period of infant dependency in turn severely restricts the mobility of the mother, so that both parents need to co-operate in child rearing. This is as true today as it was in our hunter–gatherer past; witness the difficulties encountered by single-parent families.

It seems reasonable to suppose that human sexual behaviour

may have evolved to become the bond that has made long-term heterosexual pairing possible in a species that is essentially polygynous. In the first place, we are almost unique amongst mammals in general, and primates in particular, in having suppressed the cyclical female sexual attractiveness and receptivity of oestrus, and exchanged it for a situation in which the woman is continuously sexually attractive to the man, and will consent to intercourse throughout the whole of the menstrual cycle, pregnancy and lactation, and after ovarian activity has ceased in the post-menopausal years. She also receives added gratification from sex in the form of female orgasm, which seems to be another uniquely human attribute. It is no wonder that in terms of copulation frequency, we are the sexiest of mammals; perhaps this is why man has developed the largest penis of all the primates. Women also have undergone a number of anatomical modifications. With the suppression of oestrus and the acquisition of an upright posture, there was presumably no longer any need for the perineal swelling that many other primates use to advertise an impending ovulation. Instead, the human breasts have developed as erotic organs; even Desmond Morris seems to have missed the significance of the fact that we are the only primate in which the breasts become fully developed at puberty, well in advance of the first ovulation. In all other primates, the breasts only develop towards the end of the first pregnancy, and seem to have no erotic significance (see Susu Ohno's comments in Chapter 1). Comparative anthropological studies show that a woman's breasts are regarded as highly erotic in most cultures, irrespective of whether they are normally uncovered, or concealed by clothing.

These fundamental changes in our sexual behaviour have certainly allowed us to lead a new type of social existence. Long-term pair bonding and the suppression of oestrus have enabled us to live in relative peace in large social groupings, where sexual aggression is a rare event. Maybe it was the skilful manipulation of sex for social ends that enabled us to evolve as the 'king of the primates'. But what lies in store for us as a species?

All mammals seem to have a number of inbuilt density-dependent mechanisms that enable them to regulate their rate of reproduction so that the species achieves an equilibrium with its environment. The principal reproductive variables are the time of onset of puberty, the rate of embryonic and fetal death, neonatal mortality, the duration of lactational anoestrus or amenorrhoea, and the time interval between successive births. Notice that all these factors operate in the female; this is logical, since as Peter Jewell pointed out in the preceding chapter, she has the greatest energy investment in reproduction and is therefore the limiting resource, especially in a polygynous species.

Unfortunately for civilized man, he has chosen to conquer each of these variables, so that we now no longer have any natural checks and balances on our reproductive rate. Improved nutrition in infancy and childhood has brought the age of menarche down almost as low as it will go, to just under 13. Improved obstetrical care has significantly lowered the fetal death rate. Paediatric skills have resulted in spectacular declines in neonatal mortality. Lactational amenorrhoea is virtually non-existent in developed countries, where most women no longer breast feed. It is therefore hardly surprising that the birth interval is now frequently less than a year.

The consequence of all this is that we will henceforth be completely dependent on artificial forms of contraception if we are ever to stabilize our rate of population growth; we have yet to appreciate fully the magnitude of this problem. Can man, who is genetically still a hunter–gatherer, keep pace with our rapid social evolution through the Stone Age, Bronze Age, Iron Age and now the Atomic Age? Our future as a species must remain a big question mark since we are fated to spend the rest of our existence living beyond our genetic inheritance.

Charles Darwin's views on the Origin of Species can still be read with profit and pleasure, although he did not seem to share these Malthusian forebodings about the future of our own species. Speciation has been made possible by the extraordinary degree of genetic variability that lies dormant in all mammals. This can readily be revealed by geographical isolation; the smaller the initial isolate, the greater the degree of divergence from the parental population. Mammalian reproduction is clearly such a complex process that it usually does not take long for an isolated population to develop fortuitous reproductive barriers which eventually prevent it mating back successfully to the parental stock. But maybe genetic changes could occasionally function as the initial isolating mechanism; it is tempting to suggest that some structural rearrangements of chromosomes could have allowed speciation to proceed in the absence of any geographical separation. An examination of the karyotypes of surviving species shows both surprising similarities between distant relatives, and dissimilarities between closely related individuals, thereby serving as a timely reminder of the enormous difficulties in trying to reconstruct the past from an examination of the present.

SUGGESTED FURTHER READING

Animal Species and Evolution. E. Mayr. Cambridge, Mass.; Harvard University Press (1963).

Models of speciation. M. J. D. White. *Science* **159**, 1065 (1968).

The role of chromosomal rearrangement in mammalian speciation with special reference to Cetacea and Pinnipedia. U. Arnason. *Hereditas* **70**, 113 (1972).

Meiosis in interspecific equine hybrids. II. The Przewalski horse/domestic horse hybrid (*Equus przewalskii*×*Equus caballus*). R. V. Short, A. C. Chandley, R. C. Jones and W. R. Allen. *Cytogenetics and Cell Genetics* **13**, 465 (1974).

G-band patterns, chromosomal homologies, and evolutionary relationships among wild sheep, goats, and Aoudads (Mammalia, Artiodactyla). C. F. Nadler, R. S. Hoffman and A. Woolf. *Experientia* **30**, 744 (1974).

Trisomy in the fetal backcross progeny of male and female metacentric heterozygotes of the mouse: I. A. Gropp, D. Giers and U. Kolbus. *Cytogenetics and Cell Genetics* **13**, 511 (1974).

Mammalian Hybrids. A. P. Gray. Slough; Commonwealth Agricultural Bureaux (1971).

Comparative Mammalian Cytogenetics. Ed. K. Benirschke. Berlin; Springer-Verlag (1969).

Sexual Selection and the Descent of Man 1871–1971. Ed. B. Campbell. Chicago; Aldine (1972).

The origin of species

An Atlas of Mammalian Chromosomes, vols 1–7. T. C. Hsu and K. Benirschke. Berlin; Springer-Verlag (1969–75).

Evolution at two levels in Humans and Chimpanzees. Mary-Claire King and A. C. Wilson. *Science* **188**, 107 (1975).

Chance and change in British long-tailed field mice (*Apodemus sylvaticus*). R. J. Berry. *Journal of Zoology, London* **170**, 351–66 (1973).

Disruptive selection. J. M. Thoday. *Proceedings of the Royal Society of London, Series B* **182**, 109–43 (1972).

The Origin of Species. C. Darwin, 1859. Reprint of first edition: London; Watts & Co. (1950).

Man, the changing animal. R. V. Short. In *Physiology and Genetics of Reproduction, Part A*, pp. 3–15. Ed. E. M. Coutinho & F. Fuchs. New York; Plenum (1974).

The evolution of human reproduction. R. V. Short. *Proceedings of the Royal Society of London, Series B.* In press (1976).

5 Specialization of gametes

C. R. Austin

Eggs and spermatozoa are complex cells uniquely endowed for their task of initiating the new individual. They have a most ancient lineage, for even the humblest members of plant and animal kingdoms reproduce with dissimilar gametes, and so mammalian gametes are the tried and tested products of an extremely long period of evolution. That being so, gamete characteristics should provide abundant evidence of fitness for their important role in life, including adaptation to specific environments. And behind the variations appropriate to different life styles we should expect to see the signs of genetic control, for gametes in all their parts should be determined as precisely as any other cell. Because of their great complexity we might also hope that a close study of the gametes would reveal a greater wealth of information on evolutionary and genetic influences than we could get from other cells.

Intimate acquaintance with eggs and spermatozoa fulfils these expectations to some degree, and we soon notice that the general morphology and physiology of the gametes are clearly appropriate to their task, and that distinct similarities exist between gametes of related species and differences between those of distant members. But then we run into difficulties, two of which are predominant: there are so many features displayed by gametes for which even the most ingenious explanations in terms of adaptation seem inadequate, and such an extraordinary range of variation occurs among and between species as apparently to defy justification.

If it be granted that the basic ideas concerning fitness and genetic control are correct, we must plainly consider the possibility that yet other factors contribute in some measure to the shaping of the gametes. To debate the various interactions involved we need a fairly detailed knowledge of gamete morphology and physiology. An adequate background was presented in Book 1 of this series; here we shall try to be more selective, looking particularly at features that appear anomalous and beg for acceptable explanation, while also acknowledging the existence of others that fit snugly into our theories of pattern and function.

EGGS

Overall size

Mammalian eggs vary greatly in size (Fig. 5-1) because the group includes the platypus and the spiny anteater, which have retained the oviparous mode of reproduction (features of eggs important for oviparity are discussed by Geoff Sharman in Chapter 2). Monotreme eggs are relatively enormous, being 2.5–4 cm in diameter; once laid they require to be incubated through to hatching, like reptile and bird eggs. Their size is due mainly to the yolk body and surrounding mucoid layer (or 'white'), and both regions are expended in the course of incubation for the maintenance of the embryo.

Marsupial and placental mammalian (eutherian) eggs are much smaller than monotreme eggs and they fall into a restricted size range, the smallest being represented in certain rodents such as the field vole *Microtus agrestis* (at 50 μm) and the largest in marsupials such as the native cat *Dasyrus viverrinus* (at 250 μm). The *Dasyurus* egg has a large yolk inclusion which seems adequately to account for its unusual bulk. It may indeed have more yolk than the embryo will need, for at the 2-cell stage the yolk is relegated to a separate body – like a small third blastomere (Fig. 5-2) – and during later cleavage it forms a shapeless 'yolk coagulum' in the middle of the embryo. Of course it could still function as a source of nutrient; its exclusion from the cytoplasm may not be because it is unwanted but because it would otherwise physically hamper cleavage, or because of a need to maintain certain critical surface:mass relations. Other mammalian eggs

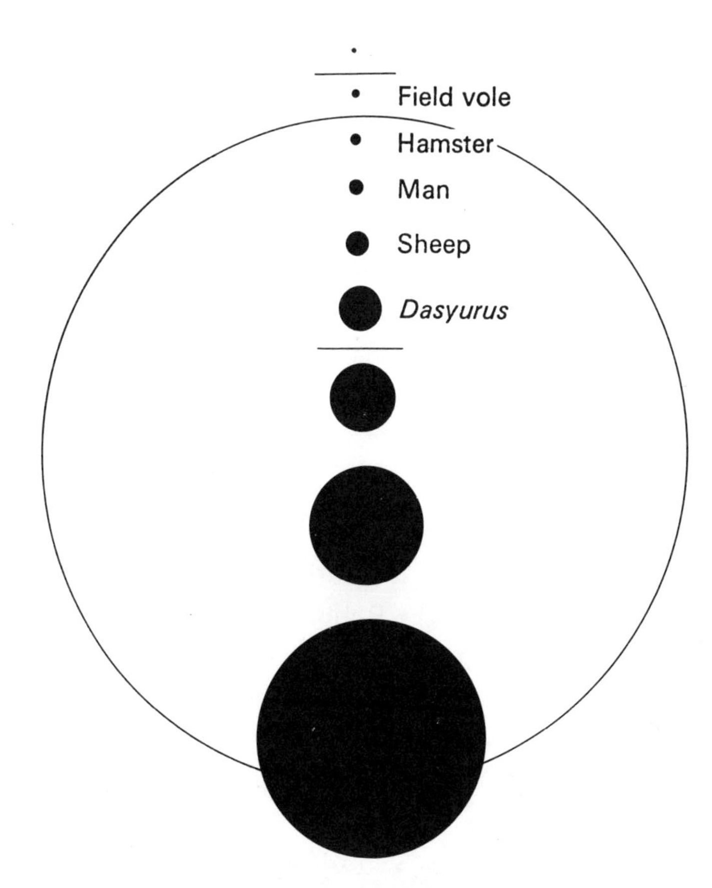

Fig. 5-1. Sizes of animal eggs. The large outline represents the yolk size in the monotreme egg and the black discs are the relative sizes of the vitellus in a wide variety of vertebrate and invertebrate eggs. Horizontal lines limit the range of marsupial and eutherian eggs, and those of many marine invertebrates. ×2.2. (From C. R. Austin. *The Mammalian Egg.* Oxford; Blackwell Scientific Publications (1961).)

also show yolk extrusion, or 'deutoplasmolysis', though in a more diffuse form; it is very noticeable in the eggs of other marsupials, such as the opossum *Didelphis* and the bandicoot *Perameles*, and has been described as a less striking phenomenon in a number of

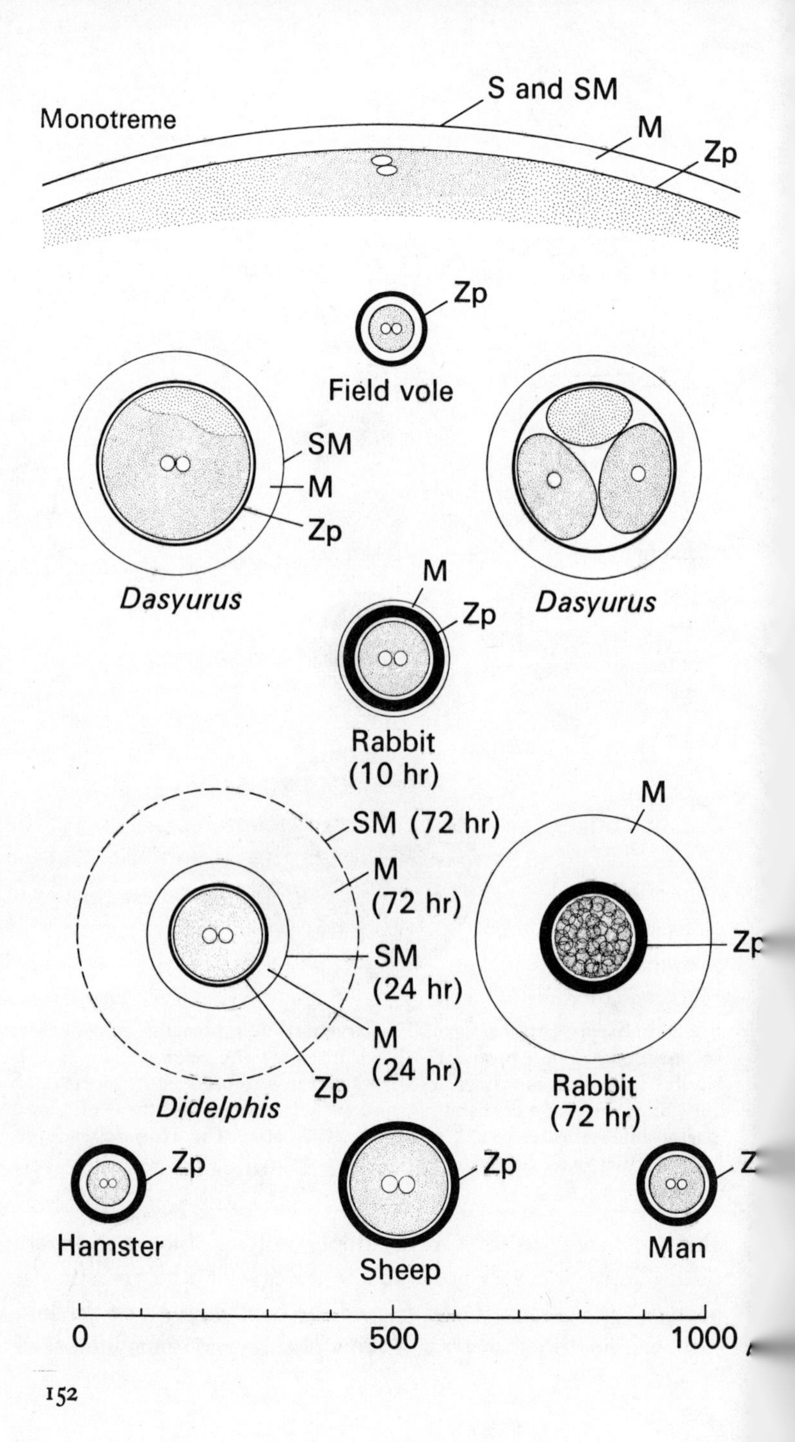

Monotreme
S and SM
M
Zp
Zp
Field vole
SM
M
Zp
Dasyurus
Dasyurus
M
Zp
Rabbit
(10 hr)
SM (72 hr)
M
(72 hr)
SM
(24 hr)
M
(24 hr)
Zp
Didelphis
M
Rabbit
(72 hr)
Zp
Hamster
Zp
Sheep
Man
0
500
1000

eutherian eggs. Among the placental mammals the egg size range is from the field vole's 50 μm to the sheep's 150 μm, and the larger eggs tend to have more droplet-like or granular yolk inclusions than do the smaller. So again egg size seems determined in part at least by the content of pabulum. The eggs of many marine invertebrates also belong in this size range (Fig. 5-1), and they too differ in their content of yolk-type inclusions – a rather striking example of parallel evolution. The biological advantage of these particular egg dimensions – for development in environments as different as the female tract and the open sea – is difficult to surmise, but may have to do with surface:mass relations and problems of adequate gaseous, electrolyte and metabolite exchanges.

Extraneous coats

Monotreme eggs display a shell and shell membrane, beyond the mucoid coat, which have much the same properties as those seen in reptile eggs, and are presumably necessary for protection during incubation. Marsupial eggs also have a shell membrane and, like hare and rabbit eggs, they acquire a mucoid coat as they pass through the oviduct (Fig. 5-2); if these coats represent some sort of needed protection, it is difficult to see why the great majority of eutherian eggs (which lack such investments) can manage without. The thin external layer remaining around the expanded rabbit blastocyst, which is believed to derive mainly from the mucoid coat, may possibly be responsible for primary adhesion at implantation, but this is uncertain. Perhaps the shell membrane and mucoid coat should be thought of as mere relics of formerly useful structures – we shall return to this point later.

Eggs of marsupials and placental mammals are released from

Fig. 5-2. Pronucleate and cleaved mammalian eggs, showing their relative sizes and the kinds of investments they display. M, mucoid coat; S and SM, shell and shell membrane; Zp, zona pellucida. Times given represent approximate hours after ovulation. (From C. R. Austin and E. C. Amoroso, 'The mammalian egg'. *Endeavour* **18**, 130 (1959).)

the ovary enclosed within a mass of follicle cells, the cumulus oophorus. Richard Blandau's fine cinematographic records provide good evidence that the cumulus is important for the egg's journey from the ovary to the site of fertilization in the ampulla of the oviduct – it offers a large surface area of suitable texture on which the oviduct cilia can obtain purchase and so move the egg along. Eggs experimentally denuded of follicle cells by enzyme action were found to be merely rotated by the cilia. After reaching the ampulla, eggs lose their cumulus oophorus, but at very different times in different species: almost immediately on arrival in the cow, sheep and opossum, but after a number of hours or even days in other species, such as the dog and cat. In view of the apparent importance of the cumulus for transport, we might expect that dog and cat eggs would reach the uterus rapidly, but in fact they take rather longer than average to do this. The early-denuded opossum egg gets there fastest, while still in the pronuclear stage of fertilization; by contrast, the other early-denuded eggs, those of the sheep and cow, come nowhere near matching the opossum. Perhaps we can best resolve this tangle by inferring that tubal contractions are more important than cilial beats for egg transport from ampulla to uterus, and that extremely rapid transport is a marsupial feature which the placental mammals have found good reason to relinquish. What the reason is remains open to debate.

In animals in which the cumulus oophorus persists about the egg, the spermatozoon must of course pass through this layer to take part in fertilization. Spermatozoa evidently depend upon the hyaluronidase they release in the acrosome reaction to digest a path through the matrix of the cumulus, and fertilization of rabbit eggs *in vitro* can in fact be prevented by treating spermatozoa with anti-hyaluronidase antibody. Oddly enough, though, the amounts of hyaluronidase carried by spermatozoa are in no way matched by differences in the dimensions or durability of the cumulus. These points are considered further, in connection with sperm characters.

The zona pellucida is a distinctive feature of mammalian eggs;

it corresponds in location and general appearance with the vitelline membrane of the avian egg, the zona radiata of the reptilian egg and the chorion of amphibian and many invertebrate eggs. Since the cytoplasmic body of the egg is limited only by a delicate plasma membrane, the need for a relatively robust non-cellular protective coat is understandably well established in the Metazoa. In the monotreme egg the zona pellucida is extremely thin; it is better developed in marsupial eggs though not nearly so well as in eutherian eggs, despite the much closer similarities in egg size in these latter two groups (Fig. 5-2). Why should placental mammals have specialized in the zona pellucida? It does seem to be a tougher structure than most of its analogues, which often have a more jelly-like consistency, and so it would constitute a better shield against adversity (though precisely against what is not clear). In addition, the zona exercises some selectivity in denying entry to nearly all foreign spermatozoa (Roger Short discusses the possible role of zona impermeability in speciation, in Chapter 4), and by means of the zona reaction it largely prevents polyspermic fertilization (discussed in Book 1, Chapter 5 and Book 2, Chapter 5). Marsupial eggs would need the same services and one infers that the thinner zona would be less effective; perhaps they depend in part on other devices. But it does not seem likely that the zona is altogether unique in exercising the functions we have considered – there appears to be a loss of sperm permeability in the chorion of the penetrated tunicate egg, for example, and the barrier to cross-fertilization in a number of invertebrate species is commonly held to be vested in this coat.

Cortical granules

Immediately beneath the plasma membrane of mammalian eggs there exist distinctive vesicular structures that correspond to bodies originally described in the sea urchin egg and given the name cortical granules. Cortical 'vesicles' would be more appropriate in view of what we know of their structure today, but the old term dies hard. Vesicular bodies of this general kind have in fact been recognized in the eggs of a strange assortment of animals – the sea urchin, the starfish, the marine worm *Nereis*, certain fish, and the frog, in addition to mammals – and the actual distribution is probably much wider. The reason for treating all these structures as homologues is that in all cases the response to entry of the fertilizing spermatozoon involves fusion of the vesicle membrane with the egg plasma membrane, and release of the vesicle contents into the space between the egg and its immediately surrounding envelope.

Consequences of cortical granule discharge vary in detail, as we might expect, but a not unreasonable inference is that in all instances these include changes inhibiting the entry of supernumerary spermatozoa. In the sea urchin, part of the material from the cortical granules joins with the chorion to form the fertilization membrane, which is impermeable to spermatozoa, while another part aids elevation of the fertilization membrane, possibly through hydrophilic attraction of water; in *Nereis* the vesicle contents greatly hydrate on release and, passing through the surrounding chorion, form a broad jelly layer on the surface of the egg which spermatozoa cannot penetrate. In fish, frog and mammals, evacuation of the cortical granules is into the perivitelline space, but the consequences are not at all clear. In mammals, one result could be a change in the properties of the zona pellucida that constitutes the zona reaction; alternatively, since the cortical granule contents have been shown to contain protease inhibitor, this could preclude further sperm entry by permeating the zona pellucida and neutralizing the sperm enzyme involved in passage through the zona pellucida. In any event, we do seem to have here the exploitation of the same kind of organelle for essentially the same intent but with wide variations in the nature of the mechanisms developed – adaptive radiation at the subcellular level.

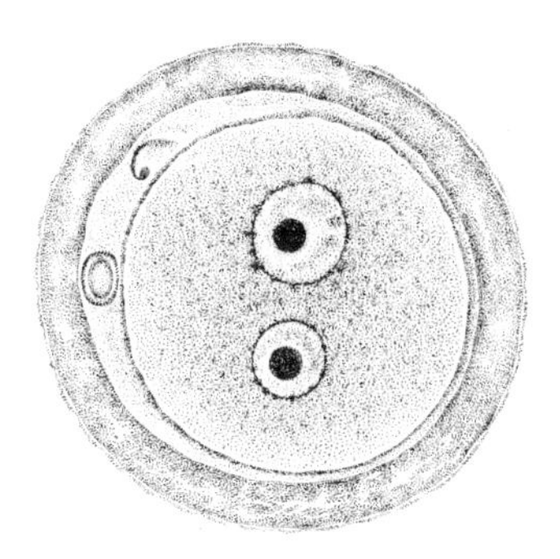

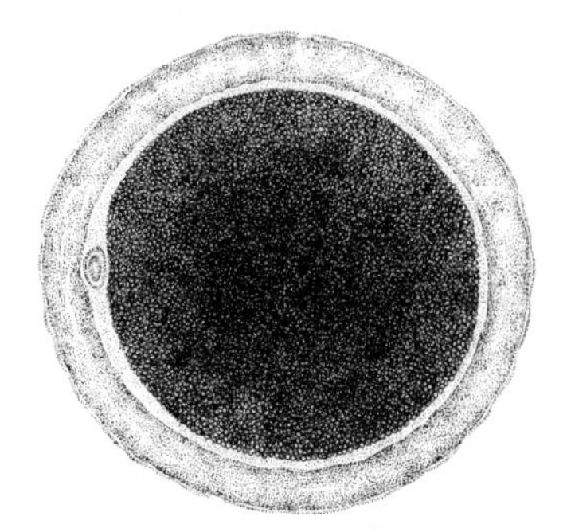

Fig. 5-3. Eggs of the Libyan jird (*left*) and ferret (*right*) contrasting the relatively clear, finely granular cytoplasm of the former with the lipid-droplet-loaded cytoplasm of the latter.

Other cytoplasmic components

Under light microscopy, whether in the fresh state or after fixation and staining, mammalian eggs differ materially in appearance to the extent that it is possible to distinguish the eggs of any one species from those of any other. Thus, in the Libyan jird the cytoplasm is remarkably transparent and homogeneous with a fine granularity (Fig. 5-3), in the rat and mouse coarser granularity is introduced and of a differing character, in the rabbit and ungulates the density of particulate material increases further, and in the eggs of the guinea pig, ferret, cat and dog there are abundant inclusions which take the form of lipid droplets (Fig. 5-3).

In addition to species differences, characterization can be made within species – Allan Braden reported that he could recognize the eggs of each of four inbred strains of mouse (C57BL, CBA, A and R3) (Fig. 5-4), and could even tell between these eggs and those of the F_1 hybrids from any two of the strains. These differences showed up only after the meiotic divisions, evidently as delayed expressions of the diploid genome – before ovulation the eggs all looked very much the same.

At the ultrastructural level further species differences become evident. There are of course similarities in the general form of the organelles, cytomembranes and microtubules; variations occur in

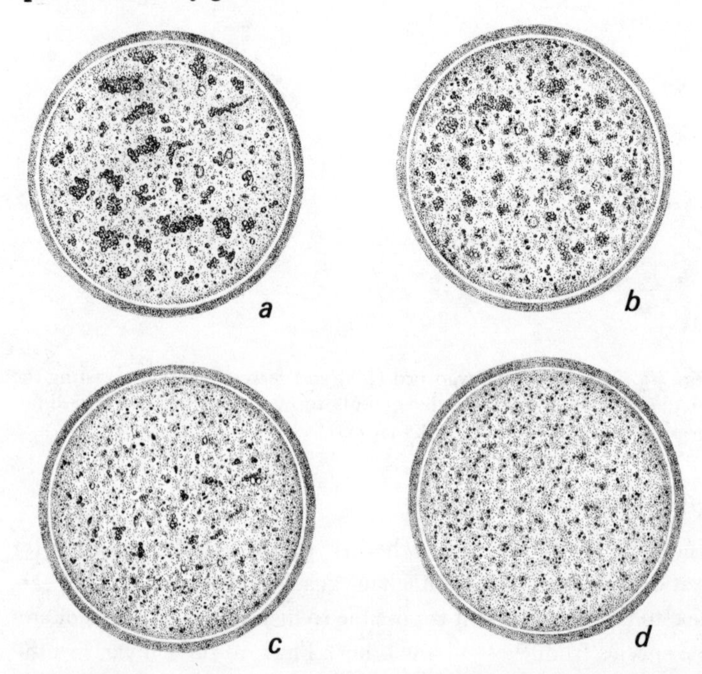

Fig. 5-4. Eggs of four inbred mouse strains: (*a*) C57BL, (*b*) CBA, (*c*) A, (*d*) R3. The cytoplasmic granularity is distinctive in each strain. (After A. W. H. Braden, 'Strain differences in the morphology of the gametes of the mouse'. *Australian Journal of Biological Science* **12**, 65, Fig. 1 (1959).)

the quantitative relations between these structures and in the properties of cytoplasmic inclusions, many of which must presumably represent stored nutrients. One is tempted to believe that there is an almost unlimited number of minutely differing ways in which eggs can be put together to achieve functional entities with similar potentials for living and developing. Probably the most striking example of adaptive radiation in the finest detail is to be found in eggs at this level of analysis. And in each instance, it seems, the mélange in all its intricacy is fully prescribed by the gene complement. Whole organisms have vast numbers of genes on which changes can be rung. Can a mere egg be equivalently endowed? We shall return to this question later.

SPERMATOZOA

Acrosome and enzymes

The acrosome is a feature of nearly all metazoan spermatozoa – the rare exceptions are in certain invertebrates and teleost fish. In shape and size, acrosomes show remarkable diversity but have long been recognized as distinctive of the species (Fig. 5-5) yet similar among closely related species. Sometimes there are also broad similarities between acrosomes in animals that are not closely related, as in the squirrel and the guinea pig (different sub-orders) and in the goat and the dog (different orders).

There is evidence in a number of species that the acrosome contains lytic and hydrolytic enzymes, which is consistent with the idea that it is a modified lysosome. The enzymes demonstrated in non-mammals (including a toad, two molluscs and three marine worms) exert a lytic action on the chorion, holes left in this coat by penetrating spermatozoa testifying to the role of the enzymes. Mammalian acrosomal hyaluronidase and a membrane-bound protease are considered to be necessary for sperm penetration through the cumulus oophorus and zona pellucida. So far the data are logical, but several other enzymes have also been identified in the mammalian acrosome, including a soluble trypsin-like protease ('acrosin'), catalase, carbonic anhydrase, lactic dehydrogenase, acid phosphatase, aryl sulphatase, β-*N*-acetyl-glucoseaminidase and phospholipase A. No function in fertilization has been convincingly assigned to any of these.

In those species in which the spermatozoa lack an acrosome, the egg does not have a chorion or else the outer investment is incomplete; in insect, cephalopod and fish eggs, sperm entry can only occur at micropyles, small apertures in the impermeable envelope.

The amounts of enzyme carried in the acrosome vary a good deal, judging by the greater ease with which they can be demonstrated in some species than in others. The guinea pig spermatozoon has about fifty times as much hyaluronidase as the

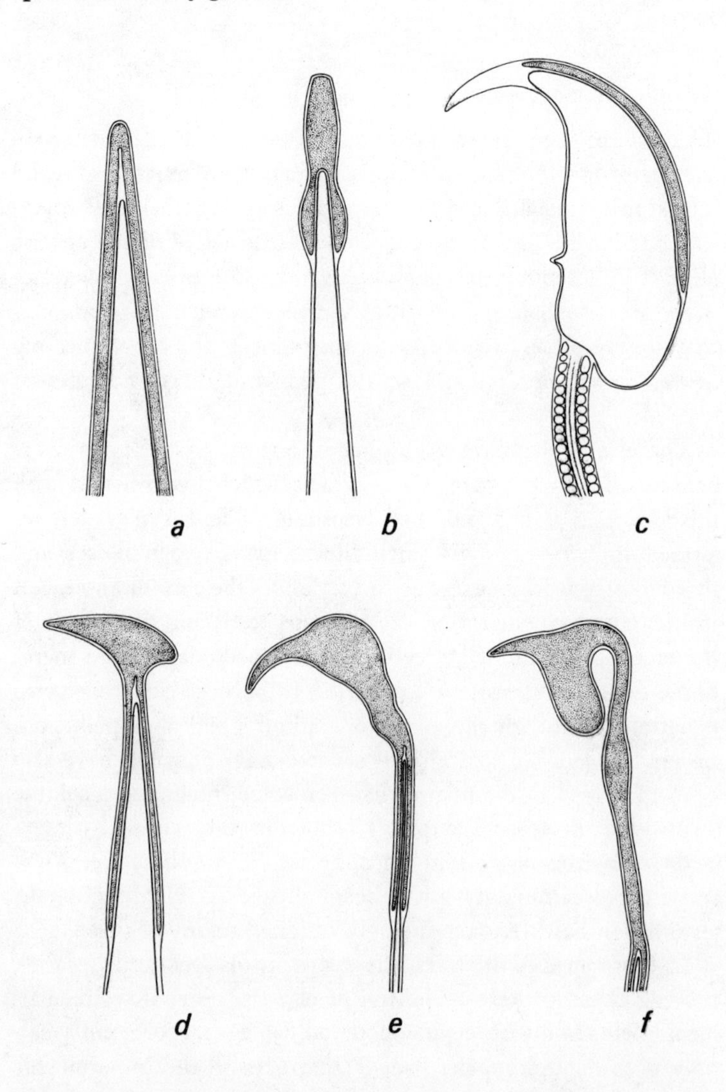

Fig. 5-5. Vertical sections of sperm heads of different mammals (*a*) Rhesus monkey, (*b*) Russian hamster, (*c*) mouse, (*d*) chinchilla, (*e*) guinea pig, (*f*) ground squirrel. The acrosome is stippled and the nucleus (only a small part of which is shown in (*e*) and (*f*)) is white. (After D. W. Fawcett, 'A comparative view of sperm ultrastructure'. *Biology of Reproduction* Supplement 2, 90, Fig. 4 (1970).)

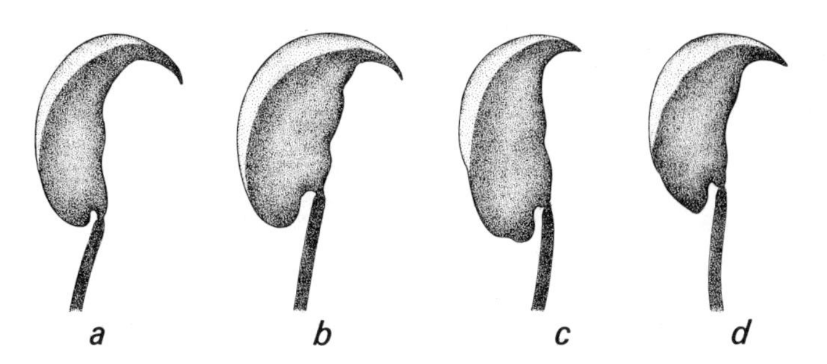

Fig. 5-6. Sperm heads of four inbred mouse strains: (*a*) C57BL, (*b*) CBA, (*c*) A, (*d*) R3. The strains can be distinguished by the slight differences in the shape of the head, especially the nuclear (dark) region. (After A. W. H. Braden, 'Strain differences in the morphology of the gametes of the mouse' *Australian Journal of Biological Science* **12**, 65, Fig. 2 (1959).)

rat spermatozoon; this comes as no surprise when we see the enormous size of the guinea pig acrosome, but the need is obscure since the cumulus oophorus surrounding the eggs of the two species is not notably different in dimensions or properties.

Nuclear shape

The outlines of the nuclei of mammalian spermatozoa differ with the species, as well as with strains within the species (Fig. 5-6). Overall the sperm nuclei range from roughly oval to spatula- or sword-shaped. In profile there is sometimes a resemblance with non-mammalian sperm nuclei, but where the mammalian sperm nucleus does appear to be quite different is in its relative flatness, cross-sectional dimensions being in the ratio of 1:4 or 1:5. By contrast most non-mammalian sperm nuclei tend to be round in cross-section. The nuclear shape in the mammalian spermatozoon is probably relevant to the manner in which the spermatozoon penetrates the relatively tough zona pellucida, the anterior part of the cell executing a side-to-side movement so that the sperm head seems to slice its way through the zona substance. In this process the protease bound to the sperm head membranes could exert a

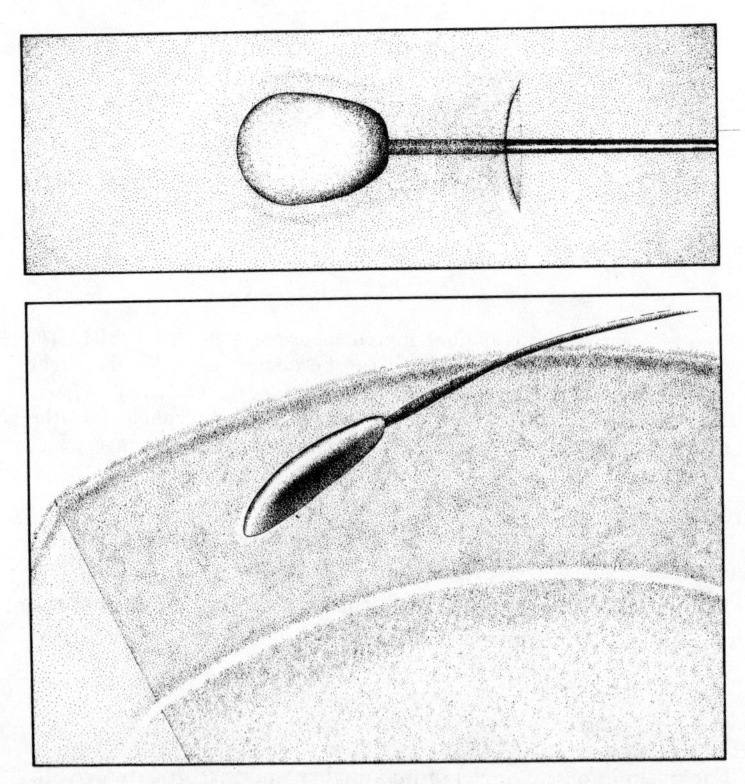

Fig. 5-7. A guinea pig spermatozoon in the act of penetrating the zona pellucida. The bulk of the acrosome has been detached in the acrosome reaction, but the form of the nucleus is unaltered. Eutherian sperm nuclei have a rigid consistency. Surface view above, sectional below.

kind of lubricant action. Penetration follows an arcuate course through the zona, with the sperm head presenting a flat surface towards the vitellus (Fig. 5-7).

If this is a true interpretation of the mechanism of mammalian sperm penetration, the effect could well depend on the substance of the sperm head being rigid rather than plastic in consistency. The plasticity of the nuclei in many non-mammalian spermatozoa is evident from the way they become drawn out as they pass through the relatively small holes produced in the chorion by the

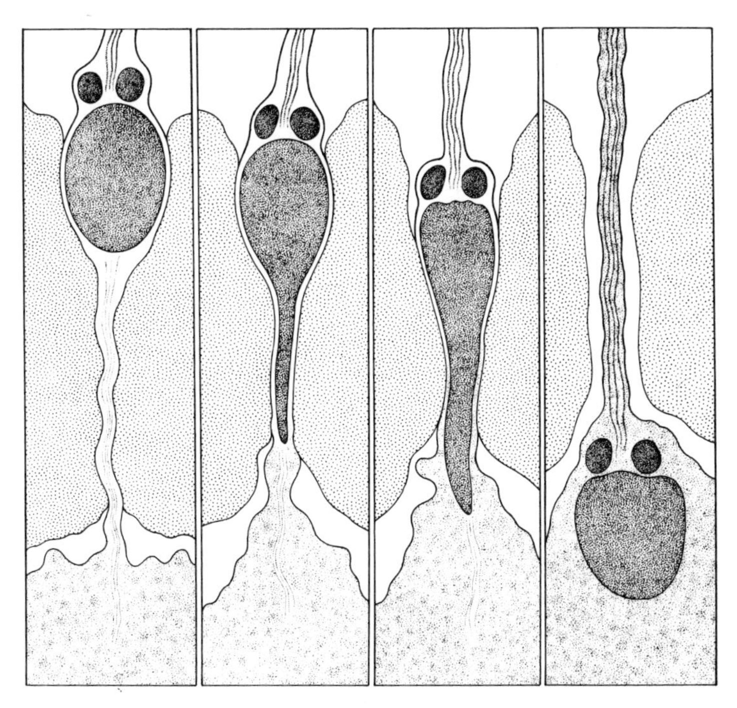

Fig. 5-8. Sagittal sections of sperm heads of the acorn worm *Saccoglossus* at stages in its penetration through the egg chorion. The nucleus (dark region) undergoes considerable distortion during this process, testifying to its plasticity. (After L. H. Colwin and A. L. Colwin, 'Role of the gamete membranes in fertilization in *Saccoglossus kowalevskii* (Enteropneusta); II Zygote formation by gamete membrane fusion'. *Journal of Cell Biology* **19**, 501–18, Fig. 1 (1963).)

acrosomal filaments and associated enzymes (Fig. 5-8). The shape of the mammalian sperm nucleus is unaltered during transit of the zona (Fig. 5-7). The difference between the two types of spermatozoon becomes evident also under many experimental circumstances, when non-mammalian sperm nuclei are found to disintegrate more readily than mammalian sperm nuclei, the latter requiring relatively powerful chemical agents to produce the same change. The rigidity of mammalian sperm nuclei has been shown by Mike Bedford to be due in part to a high degree of

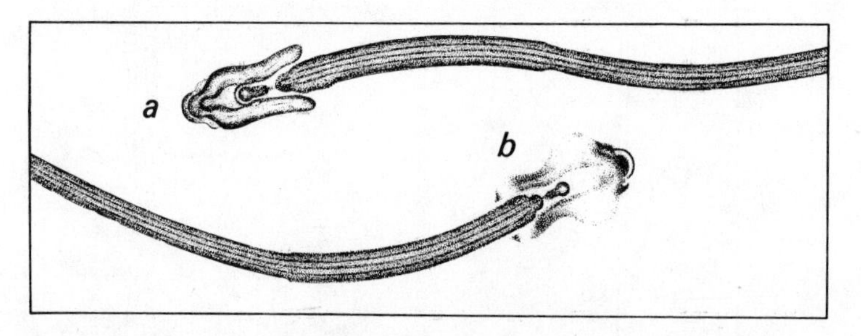

Fig. 5-9. The head of a bandicoot (*Perameles nasuta*) spermatozoon, (*a*) soon after recovery from the animal, (*b*) after being kept in buffered saline solution on a slide for about 30 minutes. The nucleus has undergone partial autolysis.

disulphide bonding between chromatin molecules. Such bonding (and rigidity) is evidently a late development in evolution for it typically exists in placental mammals; the nuclei of marsupial spermatozoa occupy an intermediate position, for they break down fairly readily without special treatment (Fig. 5-9). Possibly marsupial spermatozoa have a different method of penetrating the zona pellucida, for the heads are not so flattened in cross-section and have a curious pivoted setting on the tail (Fig. 5-10).

Nuclear envelope

Mammalian spermatozoa all show, at the posterior end of the head, a very odd structure. This is a scroll of what looks like surplus nuclear envelope (Fig. 5-11), which arises during condensation of the spermatid nucleus. The scroll is wrapped around the neck region of the sperm tail and is much greater in extent in some species, such as the bat and dormouse, than others, like the guinea pig and rat. In one animal, the bush baby *Galago senegalensis*, the complex folds of membrane are especially abundant and invest half the midpiece as well as the neck of the spermatozoon. Don Fawcett suggests that the membrane scroll might represent a store of phospholipid for endogenous meta-

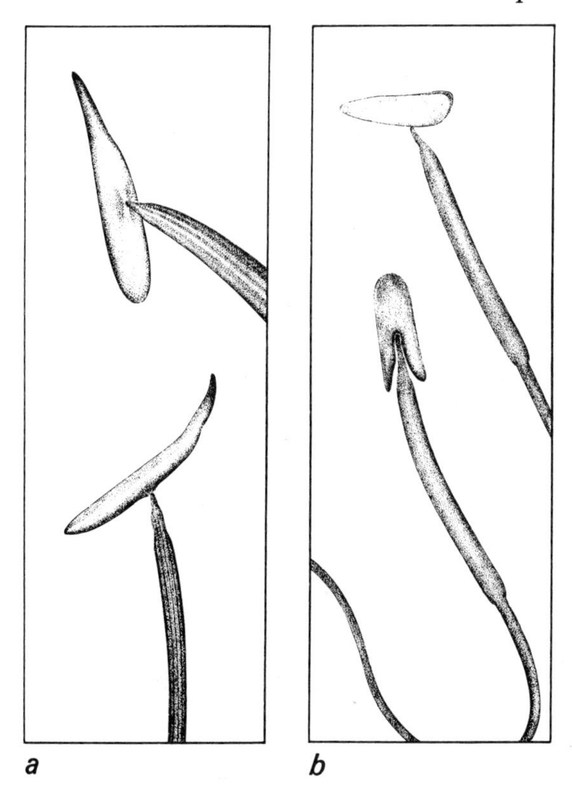

Fig. 5-10. View of the pivot-like attachment between head and tail of the spermatozoa of (*a*) the tiger cat *Dasyurops* and (*b*) opossum *Trichosurus*. The head is free to move through about 90° in the plane of the tail.

bolism (should the spermatozoon need to fall back on this process through lack of extraneous substrate). If this is correct, one is still left to account for the species differences. Alternative explanations are invited.

Centrioles

Throughout the non-mammalian metazoan series, flagellated spermatozoa with the typical axonemal complex have two distinctive centrioles which are easily recognized ultrastructurally by the

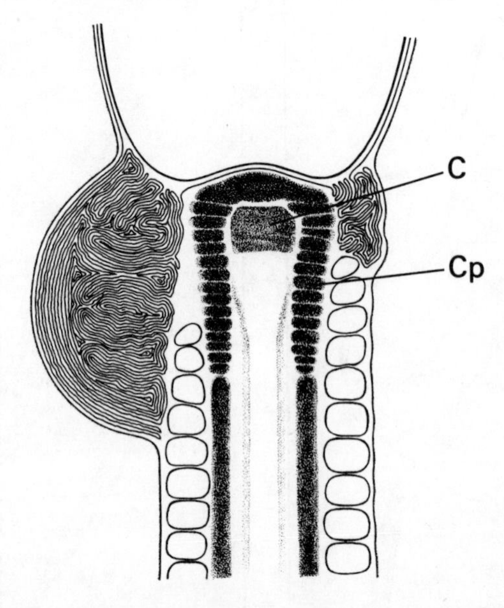

Fig. 5-11. The neck region of a Russian hamster spermatozoon seen in electron microscopic sagittal section shows a membrane scroll seemingly made up of surplus nuclear envelope. C, proximal centriole; Cp, connection piece. (After D. W. Fawcett, 'A comparative view of sperm ultrastructure'. *Biology of Reproduction* Supplement 2, 90, Fig. 7 (1970).)

ring of nine triplet microtubules. In mature mammalian spermatozoa, however (at least in those of marsupials and eutherians), things are different; as Don Fawcett points out, the distal centriolar region in the neck of the spermatozoon displays structures quite unlike standard centrioles, and indeed it is occupied largely by blocks of dense material forming the 'connecting piece' and these are associated with the anterior ends of the coarse fibres of the flagellum (Fig. 5-11). Two centrioles are clearly evident in the early spermatid, so the modification takes place during the later stages of spermiogenesis. The significance of this change is difficult to interpret. If, as is generally supposed, the distal centriole of the non-mammalian spermatozoon is charged with initiating the flagellar beat then one must conclude that in mammals the function is taken over by other structures in the neck region, presumably by the connecting piece. The anterior ends of the axonemal microtubules are imbedded in the connecting piece which may well serve as impulse initiator.

Tail structure

Mammalian spermatozoa typically have a ring of nine coarse fibres surrounding the 9+2 microtubular axonemal complex; in this they are unusual but not unique. Several non-mammalian spermatozoa (including those of certain insects, molluscs and snakes) also have the 9+9+2 arrangement, though in none of these are the coarse fibres quite so well developed as in mammalian spermatozoa. Even among the mammals the coarse fibres vary enormously in their cross-sectional area, disposition and shape (Fig. 5-12). In general, spermatozoa with larger tails tend to have larger coarse fibres.

Coarse fibres in the sperm tail probably characterize animals with internal fertilization. Some non-mammalian species with internal fertilization have spermatozoa without coarse fibres (such as the tunicates) but in these the spermatozoa are carried in currents of sea water; coarse fibres seem to be developed specifically where the suspending medium is modified from the external state and especially where, as in mammals, spermatozoa must make their way through mucous secretion in the female tract. The idea once popular was that the coarse fibres represented additional motile machinery for the spermatozoon, largely taking over the function of the axonemal complex. Indeed evidence was adduced that proteins resembling actin and myosin were present in the coarse fibres, leading to the suggestion that these structures had contractile properties analogous to those of muscle fibres. The explanation was not wholly convincing for one would then have expected spermatozoa of species in which tract secretions were unusually viscous to be better equipped. Thus in the woman and the cow, spermatozoa must pass through thick cervical mucus, apparently to a large extent by their own efforts, and might be expected to have well developed coarse fibres for this purpose; but they do not. On the contrary, the spermatozoa with the thickest coarse fibres belong to certain rodents such as the ground squirrel, Chinese hamster and guinea pig (Fig. 5-12) whose spermatozoa are carried passively right up to the uterotubal junction during the process of coitus, and so should have comparatively little swimming to do. Data obtained with later improvements in electron microscopic and analytical techniques have, however, not supported the former ideas, and the coarse fibres are now thought to be composed essentially of non-contractile structural protein. In addition Mike Bedford and his colleagues have shown that especially in eutherian spermatozoa, the molecular components of these fibres are extensively cross-linked by disulphide bonds which tend to confer rigidity. The current view then is that the coarse fibres act as stiffening devices, and consistently rodent spermatozoa do tend to have a flagellar beat of comparatively low amplitude. Evidently a stiffer tail is more appropriate for passive transport. It has been calculated too that the bending movement of bull and sea urchin spermatozoa is of the same order, provided allowance is made for the temperature difference of their normal environments, which means that the axonemal complex can quite adequately provide motility for eutherian spermatozoa.

Large variations are also seen in the dimensions of the mitochondrial aggregation that distinguishes the midpiece of the spermatozoon (Fig. 5-12 and 5-13), and also clear-cut quantitative differences apply to the number of gyres in the mitochondrial helix. Human and bull spermatozoa have some ten to twelve gyres, whereas the mouse spermatozoon can muster about ninety

---

Fig. 5-12. Electron microscopic cross-sections of the sperm tails of: (*a*) ground squirrel *Citellus lateralis*, (*b*) opossum *Didelphis marsupialis*, (*c*) bat *Myotis lucifugus*, (*d*) Chinese hamster *Cricetulus griseus*, (*e*) guinea pig *Cavia porcellus*, (*f*) Suni antelope *Nesotragus moschatus*. Striking differences exist in the overall girth of the tail, in the size and arrangement of the coarse fibres, and the width of the mitochondrial sheath. (After D. W. Fawcett, 'A comparative view of sperm ultrastructure', *Biology of Reproduction*, Supplement 2, 90, Fig. 10 (1970).)

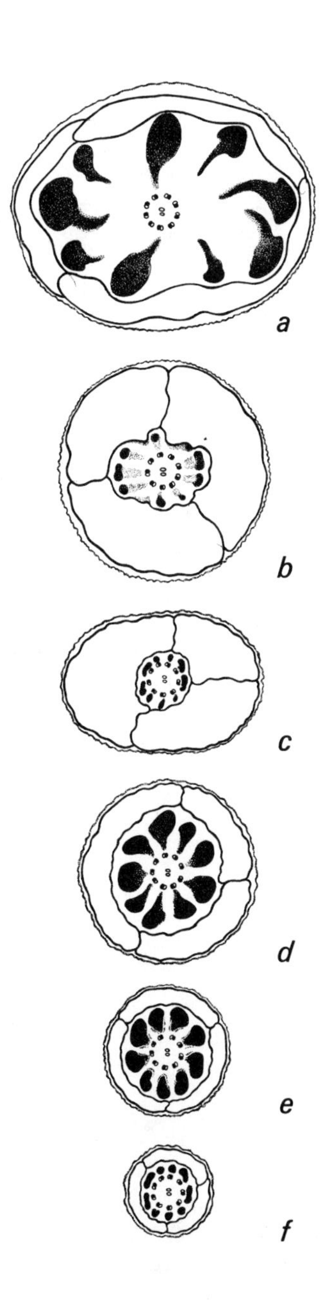
a
b
c
d
e
f

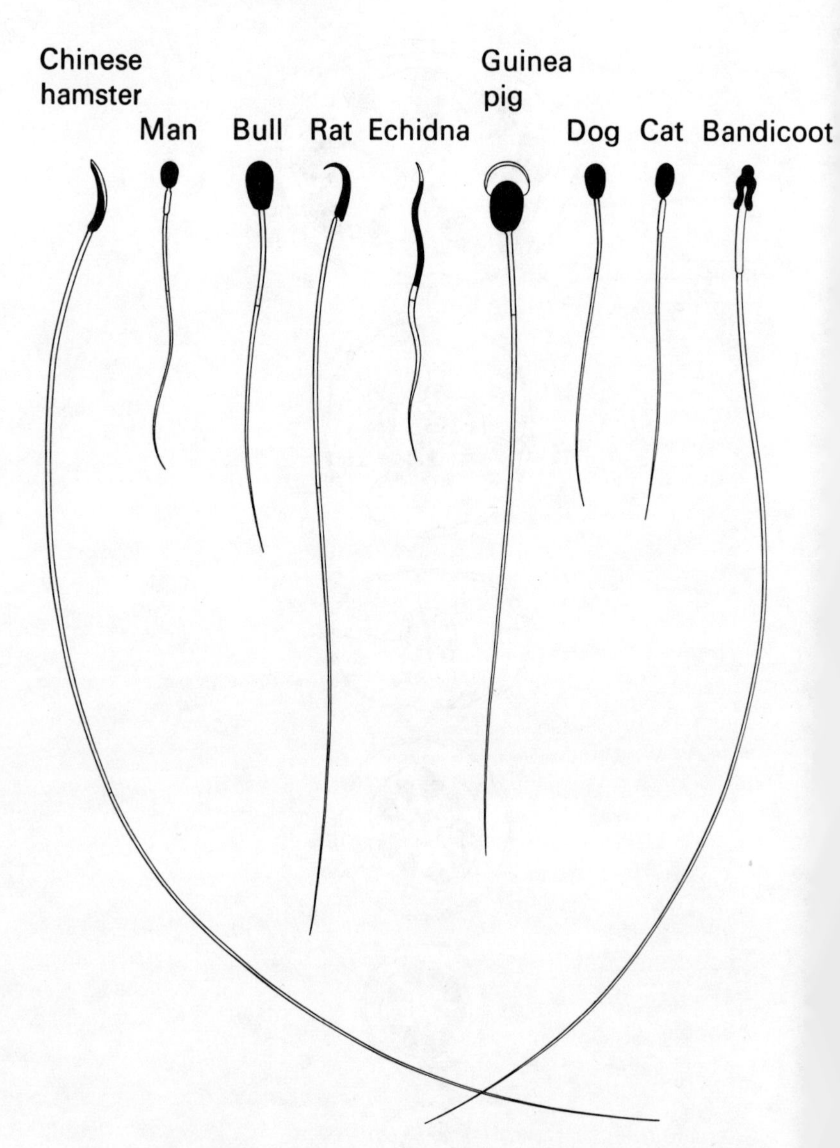

Fig. 5-13. Mammalian spermatozoa, showing differences in shape and dimensions. (From C. R. Austin. *Fertilization.* Englewood Cliffs, New Jersey; Prentice Hall (1965).)

and the rat as many as 350. The total length of the mitochondrial helix also differs considerably, occupying anything from 7 per cent of the sperm tail (as in man) to nearly 50 per cent (as in the Chinese hamster). This means that both the gyre number and the length of the midpiece are roughly inversely related to the magnitude of the swimming task with which the spermatozoa are faced in the female tract. One would surely have expected just the opposite if the mitochondria supply the energy for motility and if gyre number and length of midpiece are any kind of guide to mitochondrial activity. Evidently the relation between morphology and function is not a simple one, and this is borne out further by the fact that sperm mitochondria themselves vary greatly between mammals in individual size, shape and ultrastructure, some having far more complex systems of cristae than others. On the other hand there is something of a direct relationship between mitochondrial magnitude and coarse fibre diameters and this could be explicable on the grounds that more energy would be required for bending thicker stiffeners. But there are still some curious correlates. The Chinese hamster spermatozoon with large coarse fibres has small simple mitochondria, while the opossum goes in for small fibres and large complex mitochondria (Fig. 5-14). We obviously need more information before these structures and functions can be satisfactorily related.

Metabolism

Spermatozoa of marine organisms must survive in an inhospitable environment and so they depend on endogenous metabolism of phospholipid reserves for energy requirements; they have a limited ability to utilize external sources of energy as is evident when these are experimentally provided. Mammalian spermatozoa can profit by a rich variety of nutrient materials in the female tract secretions, but they still retain their ancestral ability to process endogenous phospholipid for energy production. The spermatozoa of different mammals show different capabilities for exogenous metabolism: for example, acetate is oxidized more

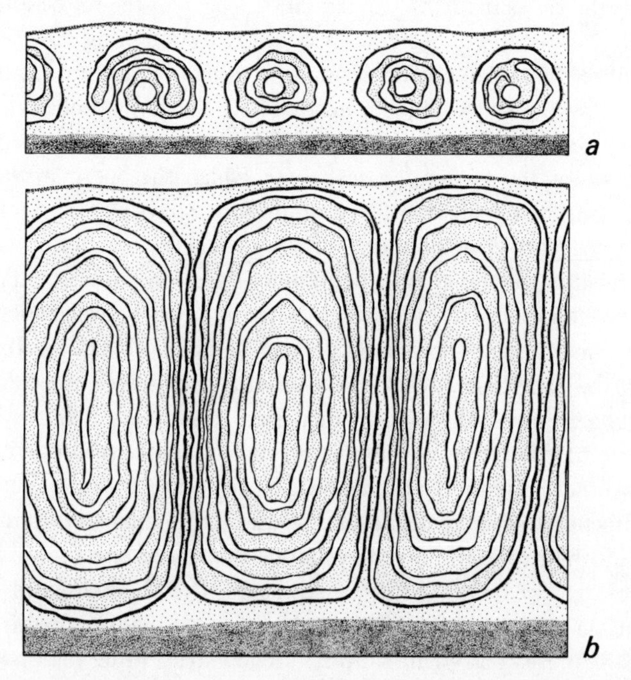

Fig. 5-14. Electron micrographic sagittal sections of part of the mitochondrial sheath of: (*a*) the late spermatid of the Chinese hamster *Cricetulus griseus*, and (*b*) the spermatozoon of the opossum *Didelphis marsupialis*. (After D. W. Fawcett, 'A comparative view of sperm ultrastructure'. *Biology of Reproduction* Supplement 2, Figs. 12 and 13 (1970).)

efficiently than glucose by ram and bull spermatozoa, but dog spermatozoa prefer glucose; sorbitol can be oxidized by ram and bull spermatozoa but not by boar and stallion spermatozoa. There are many more differences of this kind, and differences also show up in substrate concentrations in the tract secretions of different animals – all evidence of the functional flexibility referred to in the previous section.

Fertilizing capacity

That spermatozoa might vary in their innate fertilizing skill has long been suspected because of differences in fertility observed between stud males (stallions, bulls or rams). Properties of seminal plasma can have a considerable influence on the fertility of spermatozoa, but experiments involving mixed insemination with ejaculates from two or more males have revealed that the genetics of the spermatozoa themselves is often critical. Performance in mixed insemination is directly related to the fertility of the animal and also remains constant over quite long periods of time. It is difficult to offer more than mere suggestions as to the biological basis of high sperm fertility; greatly reduced fertility, on the other hand, has been linked with morphological defects of the spermatozoa arising from rather rare mutations.

PATTERNS OF GAMETE CHARACTERS

Surveying characteristics of mammalian gametes as we have just done reveals that in the main the similarities and differences between both eggs and spermatozoa reflect what can be regarded as evolutionary and genetic relationships among mammals. This has been remarked upon often before – recently by Mike Bedford following a study of the fine structure and surface properties of primate spermatozoa. But where similarities occur between distantly connected mammals, or between mammals and non-mammals, we must seek other influences responsible for shaping these cells.

Similarities between the eggs of monotremes and those of birds and oviparous reptiles have already been remarked upon and are logical enough in view of the fact that in all three groups protracted embryonic development takes place before hatching. Quantitative differences naturally exist because the young reptile must fend for itself, while the young of the other two receive parental care. Similarities also exist between the eggs of the higher mammals and those of many marine invertebrates, and this is more surprising because of the radically different environmental and other conditions. We surmised that surface:mass ratios could be of fundamental importance here; significant too must be the fact that the marine embryo becomes a free-swimming food-seeking larva at about the same stage as implantation occurs in most mammals, suggesting equivalence in the food stores initially present in the egg. So similarities can come about through the chance coincidence that similar solutions may meet very different needs, with quite unconnected genetic or environmental relations.

Sperm morphology has somewhat naturally attracted a great deal of attention and Åke Franzen and Bjorn Afzelius have come to distinguish two primary groups among the spermatozoa of invertebrates and lower vertebrates, namely the 'primitive' and the 'modified' (Fig. 5-15). 'Primitive' spermatozoa have a small rounded nucleus with an acrosome mounted on the leading surface (in those that have acrosomes), a simple flagellum consisting only of axonemal complex, and a midpiece made up of two or more mitochondria bunched in the neck region. This sort of spermatozoon is found amongst a very wide range of metazoan phyla, from the simple sponges (Parazoa) right up to bony fish, but apparently only in species that shed their spermatozoa into the surrounding water, wherein also fertilization takes place. 'Modified' spermatozoa mostly assume worm-like forms, in which the head is a long tapering cone and the midpiece (mitochondrial sheath) extends some distance down the tail, and these are found in animals that either transfer spermatozoa by intromission for internal fertilization, or release the spermatozoa near the female within gelatinous coverings. As we have seen,

Fig. 5-15. The occurrence of 'primitive' and 'modified' spermatozoa as known in most phyla of the Metazoa. 'Primitive' spermatozoa are seen typically in animals that shed gametes in open waters, while the release of 'modified' spermatozoa is directed, often by intromission or by transfer in spermatophores. (After B. A. Afzelius, 'Sperm morphology and fertilization biology'. In *Edinburgh Symposium on the Genetics of the Spermatozoon, Edinburgh, 1971*. Edited and published by R. A. Beatty and S. Gluecksohn-Waelsch. Edinburgh and New York (1972).)

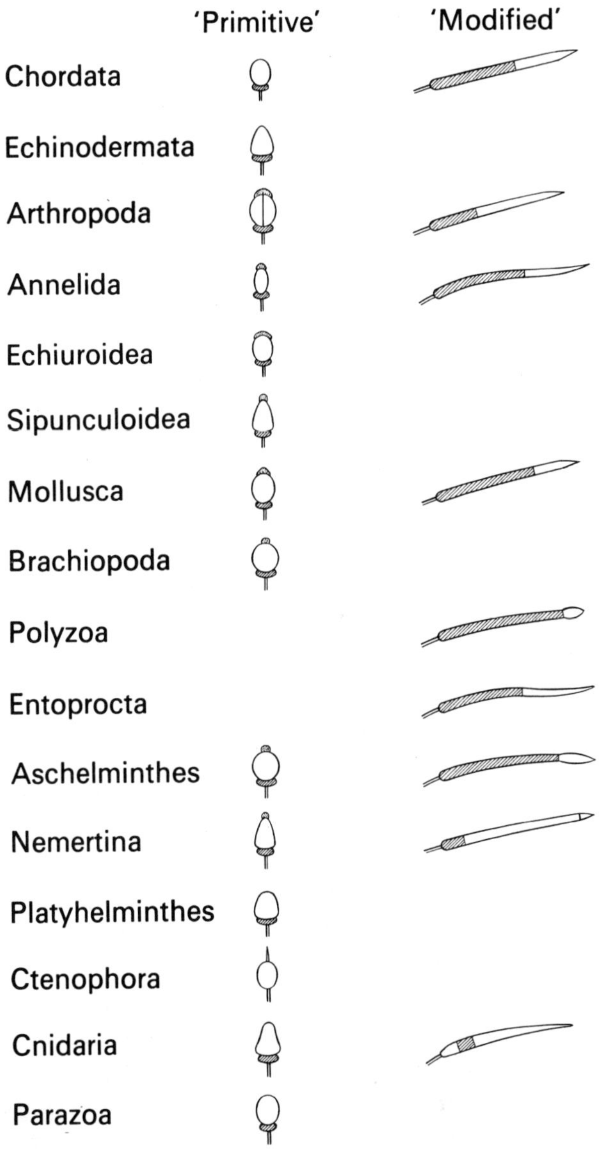

	'Primitive'	'Modified'
Chordata		
Echinodermata		
Arthropoda		
Annelida		
Echiuroidea		
Sipunculoidea		
Mollusca		
Brachiopoda		
Polyzoa		
Entoprocta		
Aschelminthes		
Nemertina		
Platyhelminthes		
Ctenophora		
Cnidaria		
Parazoa		

many of the forms involved with internal fertilization also have coarse fibres in the tail. This all seems rational enough, for one can well appreciate that a long thin (often corkscrew-like) body would pass more readily through viscous media than would a rounded form. Throughout this series, therefore, there is a much closer correlation of sperm morphology with mode of semination than with relationships between species.

When we turn to the mammals we find that the monotremes conform well with expectations for they sport the full vermiform 'modified' sperm shape. The marsupials and eutherians, however, present some puzzles; the extended mitochondrial sheath fits into the picture well but the sperm heads represent quite a new departure. Evidently additional factors have come upon the scene that override the simple needs of non-mammalian internal fertilization and it is something of a challenge to divine their nature. One factor could be the mechanism of sperm transport; we lack specific information on this process in monotremes and marsupials, but observations on several placental mammals have shown that, except in special regions like the human and ungulate cervix uteri, it depends essentially on the bulk movement of spermatozoa suspended in genital tract secretions. The main motive forces reside in the muscular contractions of the tract walls. The pressing need for the spermatozoon to possess a pointed needle- or corkscrew-like anterior extremity has largely gone, and fitness for another function becomes the focus for selection pressure. This other function – at least in eutherians – seems very likely to be the penetration of the relatively tough zona pellucida, which we saw earlier appeared to be effected by a distinctive side-to-side slicing movement of the sperm head. The tapering 'modified' type of sperm head would be a distinct liability in this sort of activity.

Assuming some validity in these arguments, we can see plenty of evidence of evolutionary development, in egg and sperm morphology and function, with a common basal pattern and the occurrence of radiation along with speciation. Also there is definite adaptive response to environmental factors expressed independently of species relationships.

From the close association of gamete morphology with species and strains within a species, we can deduce that gamete characters are genetically determined, though the degree may perhaps vary. Genetic specification can clearly be remarkably exact, for the differences between the gametes of closely related species may sometimes be limited to very small details. Also the great multiplicity of such minor points of variation is really quite astonishing, and this, as we have already noted, prompted the question – does the individual animal's genotype contain enough genes, capable of influencing gamete development, to specify all these characters? Even allowing for the multiple (pleiotropic) effects of individual genes, it would seem to be a tall order. But the available evidence firmly indicates a positive answer, and other explanations are difficult to find. As Alan Beatty remarked, having regard to the male gamete, apart from the genetic component and with allowances made for the many 'error' differences between individual cells, 'the dimensions of spermatozoa are extraordinarily independent of many biological, environmental and technical sources of variation'. Moreover, if any misgivings remain as to the adequacy of the gene complement these may be dispelled by considering the vast complexity that must be required for the specification of the battery of immunoglobulin molecules that mammals are known to be capable of producing.

Incidentally the gene differences we have been considering must very largely have been produced by point mutations rather than by chromosomal rearrangements which figure so prominently as causal factors in speciation (Chapter 4). In this connection, it is worth nothing that, by nearly all the available evidence, gamete characters are the product of *pre*meiotic gene expression. Exceptions are certain features of sperm fertility in mice which are determined by alleles at the *T*-locus whose segregation ratios point to postmeiotic (haploid) action, and perhaps also the expression of blood-group A or B antigens on spermatozoa which according to some people varies with the haploid genotype.

We might be tempted to conclude that all features of the gametes are highly heritable, and thus easily selected for, but this does not necessarily follow. Douglas Faulkner pointed out that the more closely characters of animals are associated with reproductive efficiency, the lower their heritability; thus, for example, coat colour, milk yield and conception rate in a herd of cattle investigated had heritability values (h^2) of 0.95, 0.3 and 0.01, respectively. Alan Beatty found that one measurable character of mouse spermatozoa, the cross-sectional area of the midpiece, had a much lower heritability than others (the h^2 approximating zero). We could perhaps infer from this that the bulk of the mitochondria (which largely accounts for the cross-sectional area of the midpiece) is associated in some way with reproductive efficiency (presumably in the provision of energy for sperm motility). By contrast, midpiece length was highly heritable ($h^2 = 0.76$ to 0.97, depending on method of estimation), and there was no detectable correlation with fertility – when lines of mice were selected for long and short sperm midpieces, and their spermatozoa were tested competitively in a mixed-insemination experiment, they did not differ in number of offspring sired. On the other hand, there was a positive correlation between midpiece length and body weight (in mice selected for either variable) which Beatty thought could be explained on the basis of a generalized selection for number of mitochondria (in somatic cells as well as gametes). It is interesting, though, that midpiece width varied inversely with midpiece length (the projected area of the whole midpiece remaining constant), so that differences in midpiece length might have been attributable to differences in shape of the mitochondria, rather than in number or volume, with possibly therefore no effect on the level of metabolic activity. (Other sperm dimensions have also been studied including head length, breadth and area in mouse and rabbit; all had high heritabilities, with h^2 values of 0.71 to 0.86.)

ADAPTIVE SIGNIFICANCE OF GAMETE CHARACTERS

The fact that some gamete characters are clearly meaningful in relation to function is, after all, only to be expected; what is surprising is that there are so many features of gamete morphology and physiology with no obvious biological advantage or adaptive significance. The point is brought out very clearly in comparisons between different species. Thus, why should marsupial and ruminant eggs lose their investment of follicle cells so much sooner after ovulation than rodent and rabbit eggs, and these again than carnivore eggs? Why do mammalian eggs vary so much in the nature of their yolk-like inclusions? Why does the guinea pig spermatozoon need to carry such an ornate acrosome and such an abundance of hyaluronidase? Why does the Chinese hamster spermatozoon need to be so big? Why do mammalian spermatozoa differ so radically in their metabolic capabilities? There are of course many more questions like these and they lead to the more general enigma: are all the multitudinous minutiae whereby gametes differ between species biologically advantageous in some measure in the particular environment of each species? Or would we be justified in saying rather that many, perhaps most, of the variations in gametes have no adaptive significance, are neutral to selection pressure, and, having arrived by chance mutation, will stay until by chance eliminated?

The same kind of question, but relating to parts or functions of whole organisms, has been puzzled over by geneticists for a long time. Some authorities are loth to take the 'easy way out' and accept the occurrence of non-adaptive variation, arguing that our ignorance at present is great but that in due course we shall be able to place a positive or negative value on all characters. Others equally eminent can live with the idea, at least under certain circumstances. The balance of evidence seems rather to favour this group, for non-adaptive variation is an integral part of the established phenomenon of 'random genetic drift', or the random fluctuation of gene frequencies throughout a population. The relative importance of drift and natural selection in evolution is very difficult to assess, but we do seem to be saddled firmly with the notion that characters can be widely distributed without having been selected for. If there is no significant selection against a character it could become fixed in the population, and we would thus have a uniform feature that offered no important advantage or disadvantage.

Specialization of gametes

Some people have supported another view, namely that evolution has a tendency to proceed in straight lines, as it were, so that successive generations show a character progressively changing in the same direction irrespective of selection pressure until a breaking point is reached. The process is referred to as orthogenesis and it was thought to explain why characters sometimes seem to develop to 'ridiculous' proportions, like the enormous antlers of the extinct great Irish deer (Fig. 4-1), which might well have contributed to its demise. To a less extent the lavish tail feathers of the male great-tailed grackle and the massive horns of the class IV bighorn ram (Fig. 3-1) may be similarly explicable, for they can prejudice survival; alternatively these features could have been derived through selection for success in attracting the female and this value could outway the detrimental effect for survival, as Peter Jewell points out in his chapter. Fortunately gametes are untroubled by such problems, and orthogenesis might more reasonably be invoked to account for some of their more bizarre characters. The implication of an inherent bias, however, is generally regarded with disfavour nowadays, the preferred interpretation being that natural selection prevails but the direction of genetic drift may sometimes be restricted by the nature of the character concerned or by the organization in which it exists. All the same, we should not write off the idea of orthogenesis, for the same consequence could be arrived at either through predisposition or external controls. The expression of any gene is bound to be influenced in some measure by that of others, and there could often be a kind of chain reaction, so that a single mutation causes a change that makes itself felt over a wide range of structures and functions. The drive towards adaptive fitness must involve a response of the animal or the gamete as a whole to selection pressure. Under the circumstances, it is really no less than logical that gametic (and somatic) characters of different species should differ to some degree in every little particular – without serious positive or negative influence on total fitness or adaptive value.

All this must be viewed in the context of evolution as a dynamic process – current configurations are like patterns in the shifting sands of time, and further changes must inevitably occur with further evolution. Characters are likely to come and go, and usually in a gradual manner rather than as sudden events. We should therefore see evidence of vestigial characters in gametes, namely those that were once advantageous but being so no longer are destined for elimination. Here we have another explanation for the existence of a character with negligible adaptive value.

There are several candidates for the post of vestigial character. (*a*) The mucoid coat of marsupial and rabbit eggs, and the shell membrane of marsupial eggs, could be hangovers from oviparous days. Possible functions have been suggested for these investments but other animals seem to do without them perfectly well. (*b*) Deutoplasmolysis, or yolk extrusion, by some mammalian eggs could denote a persisting excess of yolk material, important in oviparity but not really necessary for viviparity. (*c*) The range of enzymes in mammalian sperm acrosomes well exceeds known possible functions. Perhaps lysosomes are genetically prescribed more or less in their entirety, so that in exploiting the lysosome to serve as an acrosome the spermatozoon has had to take the whole package, and could do so without detriment. Enzymes once having an important role in somatic cells could well be meaningless (yet harmless) in acrosomes. (*d*) Endogenous phospholipid metabolism, though vital for such as sea urchin spermatozoa, seems unlikely to be of much value (or nuisance) to mammalian spermatozoa.

Much of the indecision in present thought on the shaping of gamete characters undoubtedly stems from our limited knowledge of their true nature and functions, but new information about these things is coming in every day and we should see a steady improvement in our understanding. Even as things are, though, we have justification for the claim that gametes offer special advantages in the study of evolution; as single cells they present problems in sharper relief than do whole organisms, and as highly complex cells they can exhibit an unusually rich array of finely differing characteristics.

Specialization of gametes

Gametes like whole organisms carry abundant evidence of their evolutionary history, and of the high degree to which all their peculiarities are genetically prescribed. Many of their characters are plainly of biological advantage, but there are also many others that lack obvious significance. Some of these are no doubt mere vestiges of structures or functions that once had real value; others may never have been of any use but confer no great handicap either. We can reasonably think of both kinds as being earmarked for removal in due course, though perhaps not for a long time hence; evolution is a steady, continuing process, not always imbued with a sense of urgency. Gametes, like Rome, are not built in a day.

SUGGESTED FURTHER READING

Edinburgh Symposium on the Genetics of the Spermatozoon. Ed. and published by R. A. Beatty and S. Gluecksohn-Waelsch. Edinburgh and New York (1972). (This book contains articles by R. A. Beatty, A. W. H. Braden and B. Afzelius, who were referred to in this chapter, in addition to other articles also relevant.)

The Functional Anatomy of the Spermatozoon. Ed. B. Afzelius. Oxford; Pergamon (1975). (Numerous articles giving a broadly comparative treatment of structure and function.)

The sperms of the British Muridae. G. F. Friend. *Quarterly Journal of Microscopical Science* **78**, 419 (1936).

Mammalian spermatozoa. M. W. H. Bishop and C. R. Austin. *Endeavour* **16**, 137 (1957).

A comparative view of sperm ultrastructure. D. W. Fawcett. *Biology of Reproduction* Supplement 2, 90 (1970).

The mammalian spermatozoon. D. W. Fawcett. *Developmental Biology* **44**, 394 (1975).

Biology of primate spermatozoa. J. M. Bedford. In *Reproductive Biology of the Primates.* Ed. W. P. Luckett. Basel; S. Karger (1974).

The Biology of the Sperm Cell. B. Bacetti and B. A. Afzelius. Basel; Karger (1976).

The Mammalian Egg. C. R. Austin. Oxford; Blackwell Scientific Publications (1961).

Fertilization. C. R. Austin. In *Concepts of Development.* Ed. J. Lash and J. R. Whittaker. Stamford; Sinauer (1974).

Fertilization. C. R. Austin. Englewood Cliffs; Prentice-Hall (1965).

Introduction to Quantitative Genetics. D. S. Falconer. Edinburgh and London; Oliver and Boyd (1964).

Index of systematic names

Index of systematic names

Subject index

Subject index

Subject index

DISCARDED
BETHANY
COLLEGE
LIBRARY